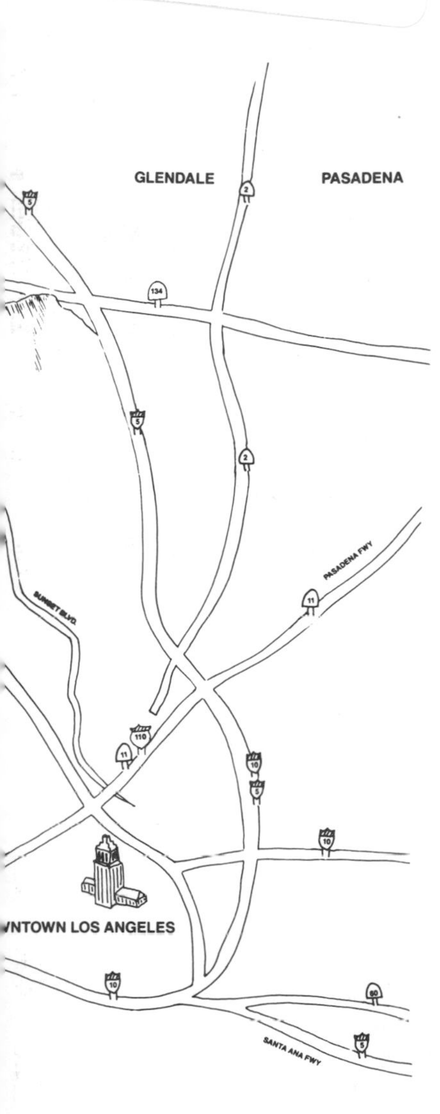

Map of Hollywood and Vicinity

1. *Optical Cinema Services,* 6649 Odessa Ave., Van Nuys, CA 91406[a]
2. *CBS/MTM Studios,* 4024 Radford Ave., Studio City, CA 91604[b]
3. *Technicolor,* 4050 Lankershim Blvd., North Hollywood, CA 91508[c]
4. *Universal Studios,* 100 Universal City Plaza, Universal City, CA 91608[b]
5. *Foto-Kem,* 2800 W. Olive Ave., Burbank, CA 91505[c]
6. *Walt Disney Studios,* 500 So. Buena Vista St., Burbank, CA 91521[b]
7. *NBC,* 3000 W. Alameda Ave., Burbank, CA 91523[d]
8. *The Burbank Studios (Columbia and Warner Brothers),* 4000 Warner Blvd., Burbank, CA 91522[b]
9. *ABC,* 4151 Prospect Ave., Hollywood, CA 90027[d]
10. *Deluxe Labs,* 1377 No. Serrano Ave., Hollywood, CA 90027[c]
11. *Paramount Pictures,* 5555 Melrose Ave., Hollywood, CA 90038[b]
12. *Raleigh Studios,* 650 No. Bronson Ave., Los Angeles, CA 90004[b]
13. *Sunset Gower Studios,* 1438 No. Gower St., Hollywood, CA 90028[b]
14. *Fox Television,* 5746 Sunset Blvd., Hollywood, CA 90028[d]
15. *Pacific Title,* 6350 Santa Monica Blvd., Hollywood, CA 90038[a]
16. *Howard A. Anderson Company,* 1016 No. Cole Ave., Hollywood, CA 90038 (see 27 for second location)[a]
17. *Consolidated Film Industries (CFI),* 959 No. Seward St., Hollywood, CA 90038[c]
18. *Hollywood Film Company,* 948 No. Seward St., Hollywood, CA 90038[c]
19. *Cinema Research,* 6860 Lexington Ave., Hollywood, CA 90038[a]
20. *Warner Hollywood Studios,* 1041 No. Formosa Ave., Hollywood, CA 90046[b]
21. *CBS,* 7800 Beverly Blvd., Los Angeles, CA 90036[d]
22. *Twentieth Century Fox,* 10201 W. Pico Blvd., Los Angeles, CA 90035[b]
23. *Culver Studios,* 9336 W. Washington Blvd., Culver City, CA 90230[b]
24. *MGM/UA,* 10000 W. Washington Blvd., Culver City, CA 90232[b]
25. *MGM Laboratories,* 10202 W. Washington Blvd., Culver City, CA 90232[c]
26. *Lorimar-Telepictures,* 10202 W. Washington Blvd., Culver City, CA 90232[b]
27. *Howard A. Anderson Company,* 3767 Overland Blvd., Culver City, CA 90034 (see 16 for Hollywood location)[a]

Names and locations are subject to change.

[a] Title and optical houses
[b] Studios and companies
[c] Labs
[d] Television networks

INTRODUCTION TO FILM EDITING

INTRODUCTION TO FILM EDITING

Bernard Balmuth, A.C.E.

Illustrations by Margot Pipkin

Focal Press
Boston • London

Focal Press is an imprint of Butterworth Publishers.

Figure credits: We are grateful to the following sources for granting permission to use figures in this book.

Figure 3-5—Big Time Picture Company, Inc.
Figures 8-1, 8-5, 9-3, 9-4, 10-1, and 16-3—Columbia Pictures, A Division of Columbia Pictures Industries, Inc.
Figures 8-2 and 8-3—Twentieth Century Fox Productions
Figures 8-6, 8-7, 8-8, and 8-9—Consolidated Film Industries
Figures 14-1, 14-3, 14-9, 14-13, and 15-1—Howard A. Anderson Company

In an effort to avoid sexism, references to gender alternate by chapter throughout this book.

Library of Congress Cataloging-in-Publication Data

Balmuth, Bernard.
Introduction to film editing/Bernard A. Balmuth; illustrations by Margot Pipkin.
p. 3 cm.
Bibliography: p.
Includes index.
ISBN 0-240-51717-2
1. Motion pictures—Editing. I. Title.
TR899.B33 1989
778.5'35—dc19 88-36609
CIP

British Library Cataloguing in Publication Data

Balmuth, Bernard A.
Introduction to film editing
1. Cinema films. Editing—Manuals
I. Title
778.5'35

ISBN 0-240-51717-2

Butterworth Publishers
80 Montvale Avenue
Stoneham, MA 02180

10 9 8 7 6 5 4 3 2 1

Printed in the United States of America

For my wife Rosa
and daughters Mary and Sharon

CONTENTS

LIST OF FIGURES

Figures

FOREWORD: WHAT IS AN EDITOR?

To anyone who thinks for a moment that an editor merely cuts film, perish the thought. Someone once said to me, ''Oh, the editor, he is the guy that cuts out the bad parts.'' At the very least, I would prefer that the person would say, ''he cuts together the good parts.''

However, he does much more than that. For example, he views the dailies every afternoon and advises the producer or the director what he needs to make the film work better. He has to make himself knowledgeable with each script and consider himself part writer, part director, and part producer. In television he has to anticipate every necessary shot knowing that the film he is working on will finish shooting in seven days. The next day the company has moved on to the next episode, and the editor is left high and dry with what he has.

The editor undoubtedly is one of the most important members of the team. Besides all of the above, he has to face the weekend knowing that Executive Producers like myself will call him at home with an urgent primal scream of, ''HELP!''

God bless the editor—and Bernie Balmuth—for writing such an informative book for all the editors of the future.

Aaron Spelling
Executive Producer

INTRODUCTION

From the days of the silent nickelodeons through the sound and color revolutions to the present, motion pictures have fascinated millions of people from young to old. To be actually associated with the business of making movies was at one time a dream that came true only for a lucky few. It was certainly not a career to be seriously pursued by most people, much less one to be included in a college curriculum.

However, the advent of television in the early 1950s brought increased opportunities for employment in the film industry and cinema courses began to proliferate. According to the American Film Institute's *Guide to College Courses in Film and Television* by 1980 7,648 cinema courses were being offered by an estimated 980 colleges with over 44,000 students pursuing degrees in the media and an additional 200,000 nonmajor students taking courses as electives. Many high schools and even some elementary schools also had begun offering their students opportunities to work on film projects.

Making a film involves many artistic and technical skills. In most cinema courses some of these skills such as writing, producing, directing, acting, cinematography, and set and costume design are taught while other skills can usually be experienced only within the production of a project. Editing is usually relegated to the latter group.

What is editing? It is "selecting the best of all angles and shots provided by the director and the best performances, and uniting them into scenes and sequences with such pacing and timing so as to portray the truth and the spirit of the script." (*The Language of the Cutting Room* by Bernard Balmuth [North Hollywood, Calif.: Rosallen Publications, 1979].) Simply stated, the primary objective of editing is to *tell the story* in the best possible way.

Editing, which scholars acknowledge to be an art, is described by Kevin Brownlow in *The Parade's Gone By* as "the hidden power" and by V.I. Pudovkin in *Film Technique* as "the foundation of film art." Pudovkin further asserts that "without editing a film is dead; with it, alive."

Since editing plays such a uniquely central role, it is indeed surprising that more schools do not have specific courses in editing that in addition to picture editing include sound, music, and trailer editing. It is even more puzzling that, so far as I have been able to determine, in many editing courses that are offered *beginning editing* is either completely ignored or is hastily glossed over.

This text attempts to alleviate this neglect by providing an indepth description of the many duties of the apprentice and assistant film editor. Their work is vital to the editor, releasing him from most technical tasks so that he can devote himself as much as possible to the artistic process of editing. These are also skills and procedures good editors have mastered. As their titles reflect, the apprentice and assistant editors are also potential editors. I will describe how their work is accomplished in the editing rooms of a major Hollywood studio or company with the understanding that these fundamental procedures may be adapted to any editing room.

Of course not everyone wants to, or can, go to Hollywood. Although most major studios are located in Hollywood, there are certainly opportunities in film editing in a wide range of productions based all over the United States and Canada. If you are in some other locale, you, the film student, having acquired an academic background, should investigate job opportunities in your area. Commercials, documentaries, industrial and educational films, and television stations offer possibilities for film or videotape editing careers. Although I refer to employment in Hollywood, many of the suggestions on job hunting discussed in Chapter 1 may be applied to your efforts in your own community.

Some of you may wish to return to academe as cinema teachers, which can be a most rewarding career for you and a benefit for your students if you precede your teaching with some solid professional experience that will equip you not only to tell them what it is really like in the filmmaking world outside the campus but also to better lead them to understand and practice creative editing and filmmaking.

I will tell you what it is really like in Hollywood, still the movie capital of the world, where the majority of major theatrical films and prime-time television is produced. *And, since most of these are still edited on 35mm film I will focus on those 35mm procedures generally practiced by the major Hollywood companies.* You will be reminded constantly that some of these techniques may vary from studio to studio and between different film labs and optical houses, but be assured that once you have learned the fundamentals as detailed in this text you should have little difficulty in adjusting to varying requirements. You must also be prepared to conform to different work demands made by various editors, producers, and directors.

For these reasons I depict throughout the book the hypothetical scenario of you, the reader, developing an editing career in the fairly structured Hollywood film industry. Thus you will begin your career as an apprentice in a major studio and eventually earn your promotion to assistant film editor. While apprenticing and assisting on features, television movies, and other film productions will not be ignored, this text will concentrate on the areas where most film entrants begin in major studios: *apprentices assigned to the shipping and receiving department and the coding room* and eventually as *assistants assigned to film editors on episodic television series.*

At this moment you may reasonably question the relevance of preparing for a *film* career when the use of *videotape* has been steadily expanding specifically in episodic series. I discuss the videotape scene fully in Chapters 1 and 17. It is sufficient to say here that in my opinion so long as film remains the major form for the longer television product and the *only* form for features, film editing basics are prerequisite for achievement in editing whether it be on film or videotape.

How might you gain entry into editing, and what are your career possibilities? Once you are hired what sort of working conditions and job security might you expect, and what will be your advancement options? How should you relate to your peers and superiors? How will your work affect your family and social life? Answers to these, and many other, questions about personal matters will be incorporated into the text as will a solid working foundation in the basics of film editing.

So now that the premise and objectives of this text have been established, welcome to Hollywood and one of the lesser known crafts of cinema—the magical world of editing!

ACKNOWLEDGMENTS

The duties of the apprentice and assistant film editor and the regulations under which they work as Guild members are indeed intricate. There are many *nuts and bolts* involved and sometimes several different procedures an apprentice or assistant might use to execute certain duties. In this text I have tried to include all the essential nuts and bolts with a generally accepted approach to getting the job done as quickly and as efficiently as possible. I required considerable help and material to explain all this clearly in a book.

For the use of illustrations I thank Susan Klos, President of Big Time Picture Company, Inc.; Columbia Pictures; Tom Ellington, President of Consolidated Film Industries; The Howard A. Anderson Company; Rosallen Publications; and Gary Gerlich, Senior Vice President, Twentieth Century Fox Productions.

I must acknowledge the advice and assistance I received at crucial periods from Robyn Guzzo, Ralph Martin, Tom McCarthy, Pat Miller, Peter Morton, Ruben Navarro, Ivy Orta and her staff, Bob Ross, Josef von Stroheim, and Lois Trent.

For their constant support and invaluable criticism of the text I am extremely grateful to Byron "Buzz" Brandt, A.C.E.; Bob Bring, A.C.E.; Michael Kahn, A.C.E.; Irving Rosenblum, A.C.E.; and Mark Tarnawsky, representative of Motion Picture and Videotape Editors Guild IATSE Local 776.

After having eagerly embraced all the above consultation, should there still exist some obfuscation and errors, I hasten to accept all responsibility for having failed to ask the right questions or, having asked them, failing to correctly interpret the answers.

I am most indebted to Arthur Schneider, A.C.E., whose steadfast and patient expertise guided me through the tortuous transition from typewriter to computer so that I might complete this book now instead of one or two years from now.

By the time this reaches publication I will have reached a decade of teaching a 35mm editing class for the apprentice and assistant film editor at U.C.L.A. Extension. I hope that this text will inspire the initiation of sorely needed similar classes at other schools. Obviously teaching such a class inspired the book. Therefore, and finally, I offer an affectionate appreciation to all of my editing students who over the past ten years taught me how rewarding teaching can be and who encouraged me to write it all down.

CHAPTER 1

JOB HUNTING AND OTHER CONSIDERATIONS

I will refer to those motion picture and television facilities that produce the most combined theatrical and prime-time television films as *major* companies. (Network prime time is that three hours of programming by ABC, CBS, NBC, and Fox generally between 8:00 and 11:00 P.M.) These companies provide the widest audience visibility, the greatest creative opportunities, the best employment possibilities, and the most lucrative salaries so it is not surprising that they are the prize targets for most editing aspirants and that the competition is most severe for obtaining beginning jobs and for continued employment thereafter.

However, before you start pounding the pavement, you should be aware of two vital facts. First, most major motion picture and television companies have bargaining agreements with unions or guilds representing the various industry professionals. You obviously must learn the rules governing employment detailed in these agreements and abide by them. Second, regardless of your professional or educational background, including experience in editing, if and when you are hired by a company covered by a union agreement, you will most likely begin at the bottom of the editorial ladder as an apprentice film editor.

THE EMPLOYMENT STRUCTURE IN HOLLYWOOD

In Hollywood, the major motion picture and television companies have a bargaining agreement with the International Alliance of Theatrical and Stage Employees (IATSE) and the Moving Picture Machine Operators of the United States and Canada (MPMO), together comprising more than two dozen West Coast studio locals. The major companies are thereby *signator* companies that may not hire you for any editorial capacity unless you are listed on the *Industry Experience Roster* that is maintained by the Producers' Contract Services Administration Trust Fund. Once listed on the roster you must become a member of the IATSE Local 776, the Motion Picture and Videotape Editors Guild. As noted, if and when you are hired in Hollywood, most likely you will begin at the bottom of the editorial ladder as an apprentice film editor—except as follows:

1. If you are working on a nonunion motion picture and your company is organized by the IATSE and becomes a signator to the Hollywood basic agreement that includes the producers' roster, you are then entered on the roster and become a member of the Editors Guild in whatever classification you might be employed, whether it be editor, music or sound editor, or apprentice or assistant.
2. If a project requires someone on its editing staff with special ability, such as a foreign language skill that you happen to have, and such a person is not available on the roster, then the signator may hire you even though you are not on the roster.

How can you get your name on that roster so that you can become employable by a signator company? You must first find employment with such a company. It seems like *Catch-22,* doesn't it? The company cannot hire you unless your name is on the roster, and you cannot get your name on the roster until you are hired! But it is not as impossible as it might seem.

A signator company may hire you, providing that all those on the experience roster in the beginning classification of apprentice editor are either working or are unemployable—the latter being those who refuse the available position for some reason, those whom the company refuses to hire for good cause, or those who simply cannot be contacted. This situation usually occurs when film production and employment are at their peaks, which is generally between August and December, although it may not be entirely restricted to this period.

When you have worked for 30 days of a 365-day period for one signator employer or for 90 days of a 365-day period for more than one employer, you can file application with Contract Services and your last employer files a verification of your employment and the quality of your work. If you are approved, your name will be entered on the roster and you then must become a member of the Editors Guild, Local 776 (Basic Agreement of Consent, Producer/IATSE and MPMO, 1982—Article IX. Preference of Employment (f), page 5).

At present there is the possibility that some changes may be negotiated in the Basic Agreement in the way that the roster is administered and how *nonunion* individuals may qualify for placement on the roster. These changes may involve the proof of your nonunion editing employment, so it is imperative that you save your paycheck stubs and any other records that will verify dates of employment, by whom employed, and your editing classification.

THE THREE Ps

The most difficult job of all is finding a job, particularly that first job. A very small percentage of the many editing aspirants will be successful in becoming union members and very few of those will make the breakthrough quickly or easily. For most of you it will be a long, hard struggle, and your principal tools will be *passion, patience,* and *perseverance.*

You should have the passion for editing and should want to make it your career more than anything else; otherwise the struggle and sacrifices may not be worth it. Do yourself a favor—before investing your time and money in looking for an editing job, either through practical experience or education, learn as much as you can about all phases of editing so as to minimize any surprises or disappointments. This text will clarify much of what you can expect as an apprentice or an assistant.

For most of you it will also take a lot of patience before you get that first assignment. In fact, it may take anywhere from a couple of months to several years. It took me ten years!

And finally, you must have perseverance. You cannot be easily discouraged or dissuaded. Despite rejection, you must continue to maintain your contacts in a diplomatic manner. (Being persistent, in other words, does not mean being a pest.)

APPRENTICE FOR HIRE?

Your primary objective in Hollywood will be to get hired as an apprentice by a signator company, hopefully for thirty days so that you will be placed on the experience roster and will become eligible for membership in the Editors Guild. The situation in other locales probably will be similar: you will need to use your own initiative to acquire regular working credentials.

Do you apply for a job at the Editors Guild? No, because its principal function is to serve and protect the members already on its roster and not to hire new applicants. Neither is it recommended that you contact studio personnel offices unless you want to apply for a clerical or unskilled job to get your foot in the door. In considering this last option, be warned that some companies, while encouraging promotions within departments, sometimes resist interdepartmental transfers of personnel.

The major studios are Columbia and Warner Brothers (combined at Burbank Studios), CBS/MTM, Disney, Lorimar-Telepictures, Paramount, 20th Century-Fox, and Universal. The major production companies are ABC Circle Films, Stephen J. Cannell, MGM-UA, MTM (Mary Tyler Moore), and Aaron Spelling Productions. You should direct your job applications specifically to the editorial department heads who are either post-production supervisors or supervising editors.

Where both feature films and television are produced, there are usually editorial heads for each area. The major studios and companies produce the majority of episodic television series as well as long-form television, which includes made-for-TV movies and miniseries, and generally provide the greatest opportunity for editorial employment.

There are also, however, smaller signator companies that produce mostly feature films or long-form TV on an intermittent basis. In these companies the person authorized to hire an apprentice might be the editor of a particular project or someone on the production staff such as the production manager or an associate producer. In addition, do not overlook as potential union employers animation companies, sound effects and music houses, and any network or cable company that employs editors from local 776 or any other union.

The first step in your search for that union apprentice job is to send résumés to prospective employers. You must obtain addresses, phone numbers, the correct spelling of the names of those who do the hiring. The first two items you can get out of phone directories or from specific industry directories such as the *Pacific Coast Studio Directory,* which is published quarterly. As for the exact names and their spelling you will have to do a lot of telephoning, but they can easily be obtained from corporate telephone operators and secretaries.

On your résumé list your cinema background and any practical experience you have had in a cutting room. Some things, however, are best left unsaid. Postgraduate degrees, impressive film awards, or editing credits may suggest that you are overqualified for an entry level position. You should emphasize your sincere desire to become an apprentice for as long as is required, your passion for cinema, and your firm determination to make editing your profession. Be certain to include on your résumé your address and telephone number, and, if these should change, send a brief reminder giving the changes.

Do not be verbose. Try to edit your résumé without omitting anything important. Have it typewritten or printed, of course, and try to limit it to one page; busy executives have a short attention span for long résumés. You should attach letters of introduction but generally not letters of reference. Mention that references can be provided if requested, and, if you have the space, you might list them individually by name and title. Indicate your hope that a meeting might be arranged at the earliest convenience. Be certain to spell names correctly and include any titles your prospective employers might have. (Several post-production supervisors, for instance, are Vice Presidents in Charge of Post-Production.)

Do not be too shocked if you fail to receive an immediate response. Remember that these people are constantly bombarded with résumés such as yours, as well as those from union members in all job classifications. If you have not heard anything in a couple of weeks, follow through with a phone call and try to arrange a meeting. You will probably have to deal with a secretary, and on your first call you should learn his or her name so that in future conversations you will be able to treat the person with the proper respect. Your best chance of getting an appointment with a major studio department head will come between April and July when production is at a low ebb. Conversely this is the most unlikely time for a new applicant to find a union job since Guild unemployment is generally at its highest.

Even though you may not be granted an interview, your résumé and your occasional telephone conversations with the secretary will familiarize the company with your name, impressing the people you talk to with your interest and determination and laying the groundwork for any future possibilities.

NEPOTISM AND OTHER CONTACTS

Any contacts you have who can open doors that will lead to that first job are more important than résumés or educational background or even previous experience. Although it is commonly and unjustly considered to be pervasive only in the motion picture industry, nepotism may be present in any industry, corporation, or small business concern. For good or bad, however, it is still the easiest way of obtaining contacts and/or jobs. Should your mother, father, or another relative succeed in getting you employment, it would be termed an act of nepotism, but it is also an act of love.

Some of you fortunate enough to be in this position may have the noble determination to make it on your own. If you persist, most of you will see other, less noble individuals walk away with available jobs, leaving you sitting outside the studio gates. The best suggestion is to take the helping hand and take the job, if you have the opportunity.

I assure you, however, that *most* editorial assignments, beginning or otherwise, are gained through some sort of contact or reference from friends, former employers, fellow workers, or especially fellow cinema students—all who form a vital nucleus for finding and obtaining employment. Do not be reticent about asking for whatever assistance anyone can give you. In return at some future date you may be in a position to help either those who helped you or others.

THE NONUNION SCENE

Now that you have taken all the initial steps in applying for a union job, what are your options in the meantime? You must realize that "meantime" may be a long, long wait and that the majority of aspirants never become union members. So you must carefully consider your alternatives, which will be dependent upon your financial situation.

If you can afford it, there may be worthwhile film and videotape courses offered at colleges and film schools in your area. Before enrolling, try to investigate the courses and instructors to make certain they are what you want and need. There are many good courses in the Los Angeles area, and because of the accessibility of the studios, many of the instructors and guest speakers—particularly in the university extension divisions—are renowned professionals.

Often, full-time cinema students in these schools welcome outside volunteers to work in production or as performers in their class projects. Should you volunteer to work on one of these projects, you will be getting some practical experience in an editing room, usually working with 16mm film. These projects have very limited budgets, so no salaries are paid, yet you will be expected to perform services as though you were on a salaried editing job whether it be on a full-time or a part-time basis. Besides the experience, you will be making new friends who may become important contacts at some future date. It is obvious, therefore, that, when you agree to work on such a project, you should be reliable, conscientious, and efficient. Such experience can then serve as a valuable addition to your résumé.

Another alternative is to apply for work in some of the many nonunion, independent companies that have recently been established. In Hollywood such companies produce a large portion of the Hollywood film output and embrace every phase of filmmaking in film and videotape. Their products and services include features, TV variety shows or specials, commercials, documentaries, educational and industrial films, animation, and various other editorial services.

Many people, unable to break into union work, may find film careers here. While contacts are always important, particularly for continuity of employment, previous experience will usually be the major consideration in getting that first nonunion assignment. The reason for this is that these companies are not as structured as the major studios and often do not budget for apprentices but instead combine apprentice duties with those of the assistant or accept volunteer help from those eager for the work experience. Nor are there any hiring restrictions, so your entry level can be that of assistant or editor—depending on whatever ability you can convince a company you have. Therefore your résumés to nonunion facilities may differ considerably from those you submit to signator companies.

An apprentice job might be available at an editorial *house* offering complete post-production services, or one only offering sound or music editing. Also, the most active nonunion production companies (such as Cannon Films, Concorde/New Horizon Corp., New World Pictures, and Vestron Pictures) usually have in-house editorial staffs that employ apprentices.

Since there do not seem to be any hiring restrictions such as an *Industry Experience Roster* in these companies, and, better still, since previous educational and practical experience seem to be of more consequence than contacts, why then do we assume that nonunion work would be a secondary choice? As noted at the beginning of this chapter the major *union* companies, having greater resources at their disposal, generally are able to employ the most qualified filmmakers and technical crews. It is assumed that your first consideration would be to work with top craftsmen on the best quality product with the most visibility. It is marvelous that this is also accompanied by good salaries and such fringe benefits as vacation and holiday pay, a pension plan, and excellent medical insurance.

Nonunion salaries, depending upon the company and its budget, can vary from nothing at all to nearly as much as base union wages. Rarely are any of the above-mentioned benefits provided, however. In lieu of salary you may be offered deferred payment or a flat sum or a certain percentage should the film be profitable, but the majority of salaries paid in nonunion companies are inferior to union wages for comparable work.

Why might you want to work for *nothing at all* or for monies you may never receive? One reason is that, in Hollywood or areas with a similar union structure, there is always that possibility that the company you are working for may become a signator, giving you an automatic entry into the union.

More important, nonunion work serves as needed experience and references for future work. Although the union companies generally produce higher quality shows, there may be some excellent creative opportunities in nonunion productions. There may be situations where, because of the less structured environment, you get the chance to do more advanced work. Conversely you may not be compensated appropriately. In addition, because they are not bound by union regulations, some of these companies may induce you to perform menial services beyond what you were hired for and to work excessive hours on a continual basis. The quality of nonunion work thus can be an unknown, with possible benefits and some serious drawbacks. Its viability will ultimately depend on the company, the people involved, the production itself, and your own career situation.

Budget problems, production delays, or unexpected release dates can force both union and nonunion editorial personnel into working long hours. A 10- to 14-hour workday five or six days a week is not uncommon in the industry. Sometimes hiring extra help will relieve the pressure and reduce the hours, but it is not always feasible. When such a situation arises, everyone has to pitch in and get the project completed on time.

Only *you* can judge when the demands on your time become excessive. If the situation is the result of poor management, ego, or temperaments, and could have been avoided, or if the demands are simply unnecessary, then you might consider objecting or refusing to be taken advantage of. On the other hand, you have to understand and accept an unavoidable emergency situation.

It can be an awkward, no-win situation. Remember that even though you may have become important to the project, no one is irreplaceable. There are many eager substitutes waiting for the phone to ring. Also, a vindictive employer can send out negative information about you and word gets around amazingly fast in a close-knit industry.

Whether you start as a nonunion or union employee, you will soon learn that the film business offers erratic employment. In either case, your best opportunity for fairly steady work is on the editorial staff of a major company, an editorial service company, or one of the permanent specialized companies.

Many of you will not be able to afford the luxury of continuing your education or working at volunteer or low paying jobs. Even if you are a union member, you will have periods of unemployment and may have to find work outside the industry to put food on the table and a roof over your head. You might consider occupations that will give you the flexibility of being able to return whenever you are out of work such as temporary clerical or secretarial work, selling real estate, or bartending. I know several union members who can rely on working in a relative or friend's establishment such as a restaurant, furniture store, or bakery. One of my former students who is a good handyman receives free rent for managing an apartment building, an arrangement he can handle even when assisting.

At the same time, without endangering your job, you will want to maintain any contacts you may have made. Doing this and exploring new possibilities and going for interviews will be a problem unless your work schedule allows you some free hours in the morning or afternoon. If it doesn't, obviously you will be at a disadvantage in competing for an entry-level position.

THE VIDEOTAPE SCENE

At present the motion picture industry is in the throes of a film/tape revolution. Electronic editing facilities have mushroomed throughout major production areas; some are flourishing while others fight for existence, in some cases because of the use of already outdated equipment.

Videotape editing equipment can be very expensive especially compared to film equipment. Also technological improvements have come so rapidly that much equipment becomes inferior or obsolete in the very competitive market even before the initial costs of the equipment have been recovered.

There are many nonunion videotape facilities, as well as union ones. Since early videotape editing was accomplished by technicians or tape operators without the benefit of any previous editing experience, there are contracts with some independent companies and networks that are shared by several technical unions. For example, NABET (National Association of Broadcast Employees and Technicians) Local 53 contracts film and videotape for NBC, while Local 57 contracts only videotape at ABC for whom film is contracted by Editors Guild 776. Also, IBEW (International Brotherhood of Electrical Workers) Local 45 contracts film and videotape for CBS.

Video products were originally used mostly in such specialized areas as commercials, documentaries, industrial and educational films, and music videos while prime-time material produced on video was primarily limited to variety programs and a few half-hour comedy series. In the last few years, however, video productions have expanded to include a large number of one-hour series such as *Knotts Landing, Dallas, Falcon Crest,* and *The Paper Chase* as well as some long-form programs such as TV movies and miniseries. At present more than a dozen one-hour, episodic series that were previously shot and edited on film are still being shot on film but are now being edited on videotape.

The majority of those who are presently editing these prime-time shows are not, as one might expect, nonunion editors with extensive videotape background nor are they videotape editors from the aforementioned technical unions but rather Local 776 *film* editors who were granted the time for on-the-job instruction by their producers. The obvious reason for this is that the knowledge and feel for editing is more important to both producers and directors than is the ability to master any editing tool, which can be learned in reasonable time if given adequate instruction.

While film editors are learning to edit on videotape equipment it is paradoxical that some videotape editors who are lacking any film background are seeking training and experience in *basic* film editing. This is because these videotape editors have had difficulty getting assignments in more preferred areas of editing. For example, some students of my basic editing class at UCLA Extension have forsaken their tape editing of documentaries and variety specials when they have gotten the opportunity to work as film apprentices or assistants. By so doing, they hope that eventually they will be considered as either film or videotape editors on the more desired projects.

I believe the assistant/editor relationship often suffers in videotape. Unlike film, with its graduated training and learning structure from apprentice to assistant to film editor, tape editing does not always offer similar entry or experience opportunities. In many nonunion situations, where the videotape editor is forced to do all the work, apprentices or assistants may not even be hired or when one is hired she may have to service more than one editor. Therefore, there are far less entry-level opportunities than in film. Even union assistants may find themselves in situations that restrict their involvement with their editors, including screenings with directors and producers, and consequently their ability to learn editing. I know of one assistant who worked on an elementary tape machine on the third floor of a building while the editor was esconced on the first floor. An apprentice served as a messenger between the two rooms.

The assistant's duties in videotape editing consist primarily of logging and identifying dailies (the previous day's shooting). There may be other duties depending on company procedure and the videotape system being used by the editor. For instance, on one system, the Ediflex, the assistant is required to *mimic,* or *compose script mimic,* which is to number each line of dialogue or piece of action in the script and make a corresponding tape reference.

How can the beginner get videotape background with costly equipment? There are tape editing classes offered, but they are quite expensive, ranging from $250. to $600. for one all-day meeting and from $600. to $1,000. for ten three-hour sessions. Their value versus this cost is questionable unless the student is already working in a television station or tape facility where editing equipment is accessible for practicing what has been learned or if there is a position awaiting the student upon completion of the course. Otherwise, it would be difficult to retain that which has been learned. Another question is, what system are you going to be taught on and what are the chances that the identical system will be used at your first job? There are about a half a dozen different systems being used on the prime-time projects and other systems on other programming. Each is sufficiently different from the other so that no matter how proficient you become on one, you will require a certain amount of instruction on another. Before selecting a course, you must also consider whether that equipment may soon be phased out of the industry by the competition or by new technology. As with film courses some colleges welcome nonstudents as volunteers on their videotape projects. You just might be accepted out of desperation, especially if you have any film editing experience.

Having a film background thus is considered an advantage for videotape editing. A final consideration is that the pay scale for most union or nonunion tape editors is much lower than that for union film editors who can usually maintain their customary salaries in either medium. For all these reasons, one must conclude that if your objective is to be a union videotape editor who might someday be assigned to a prime-time series or TV movie, gaining the many creative opportunities and economic benefits, you should first try to become a *film* apprentice, then a *film* assistant and then, preferably, a *film* editor before you become a *videotape* editor. This is one of the reasons this text has been written as an introduction to *film* editing.

MORE HINTS FOR THE HUNTER

By now it must be apparent that it will be no easy task finding an editing job in Hollywood or anywhere else. Here are a few more suggestions for the beginner.

Reside in a Location Convenient to Most Studios

Los Angeles is a sprawling community of about 465 square miles. From Walt Disney Studios northeast in Burbank to Lorimar-Telepictures (formerly MGM) southwest in

Culver City it is nearly nineteen traveling miles. Within an imaginary square of 88 miles, using West Hollywood and an adjoining part of Beverly Hills as an aproximate center, most of the major studios, film companies, film and videotape editing facilities, and network stations are located. There are, of course, more independent companies scattered throughout the fringes and outskirts of Los Angeles. (See the map inside the front cover of this book.) Similarly, production companies in other cities are often some distance from each other, or, if they are closer together, other factors, such as big city traffic, amplify the distance.

What is a "convenient" location? There are some primary considerations such as the affordability of the kind of residence you want or need and the requirements of a roommate, spouse, or children. You may get a job 5 minutes from home, but your next project may be across town, a drive of 45 minutes in heavy traffic. (Add to that 1½ hours per day and possibly another ½ hour or more in rainy weather!) Your initial reaction may be that such travel time does not present any problem for you. But be aware that this traveling is added to a day in a cutting room that is at least 10 hours and often is 14 to 16 hours.

If you are a stranger to Los Angeles or the area in which you will work, familiarize yourself with the area before you establish a permanent residence.

Have a Car

Since editing usually requires long, variable, and often unusual hours and since each new position you acquire is often at a new location, you can not depend on public transportation. Besides, many of the jobs will require the use of a car. Although owning a car in Los Angeles is crucial, there are, to be sure, some places such as New York City where owning a car may be less important or even impractical.

Present a Good Appearance

Casual attire is the mode in the editing profession and is acceptable for interviews or on the job as long as it is *neat* and *clean.* Excessive make-up, flagrant hair styling, or unkempt beards or moustaches may foster prejudice in some employers.

Have a Telephone Answering Machine with Remote Control

When a position suddenly and unexpectedly opens up, an impatient post-production supervisor may offer it to another applicant if you cannot be contacted quickly. And here's a cautionary note: Do not make your machine message a big production. Your friends may find it clever or cute but an impatient prospective employer may be irritated by it. He wants to hear your name or phone number clearly so that he can be assured he has reached the correct person. Be certain to give him the courtesy of a return call as soon as possible—even if, for some reason, you are no longer interested in the job.

Check the Trades

The daily trade papers list production charts once a week including a schedule of future films with tentative shooting dates. In Hollywood, presently, "Pictures in Preparation'' is in Tuesday's *Hollywood Reporter* while *Daily Variety* has ''Films in the Future'' on Fridays and, on various days, ''Film and TV Casting News,'' which can also be useful. (These days are subject to change.) The weekly *DramaLogue* may also give you good leads. As you become more familiar with the studios and production companies, you will be able to tell which productions are nonunion. A phone number is usually included in the listings and persistent inquiries may get you through to the editor, production manager, or another individual who will talk to you about editorial work.

Set Yourself Some Deadlines

It will be difficult to establish for yourself an exact date on which you must give up on a goal, a dream. But only a small number of you will be able to establish editing careers either nonunion or union. At certain stages in your quest you must be realistic about your future prospects.

You must determine for yourself how much time you can afford to spend on classes or on volunteer jobs, and then you will have to find some sort of interim work while you continue your efforts to get a paying editing job.

Should you start working on nonunion jobs, you must realize that employment is insecure and erratic—unless you are assigned to the in-house editorial staff of an editorial services facility or on-going production company. These latter positions would be as permanent as you will be able to find in nonunion situations. If you do not obtain one of these fairly secure positions and your jobs are few and far between, at some deadline you should start reconsidering your options.

As was stated before, getting a job in a major studio in another department is not a very likely entry into the editorial department. However, in some companies it could possibly turn out to be *a foot in the door,* so give careful consideration to any job offer, even though it may not be exactly what you want. Any temporary occupation should be performed as conscientiously and efficiently as possible. You might need it later as an employment and character reference. Also, at some deadline you may find yourself seriously considering the temporary job's career potential, should you not get into editing.

THE BIG TIME!

For some of you that lucky break may come quickly, and for others it may take several years. But a small number of you will finally get that phone call that will result in your first apprentice editing job with a signator company. As I explained earlier, in Hollywood after you have worked for a single employer for at least 30 days within a 365-day period or for more than one employer for 90 days within the same period, your employer should notify Contract Services Administration to add your name to the Industry Experience Roster and you must then apply for membership in the Editors Guild.

Patience and *perseverance* have brought you to the big time. You may have had to make certain sacrifices along the way, but you still have the *passion* for wanting to work on the editing of motion pictures so you believe it was worth the struggle and now you think you have it made. But have you?

Be warned that you have only stepped onto the first rung of a very unsteady ladder. In the motion picture industry steady employment is generally uncertain and sometimes infrequent. Some editing jobs may not be as pleasant or rewarding as you would want. You will still be in that struggle for survival, so don't throw away those three *P*s.

On the other hand, you will be a member of a select group of craftsmen contributing to the telling of a story to a vast and unseen audience. And within every job, good or bad, lies the opportunity to learn and grow in your ability as an editor. I hope you will find editing exciting and rewarding. Each rung of that ladder you ascend represents another episode in that dream that propelled you into this adventure, and there are no limits to success in the motion picture industry.

If you are not disheartened and are still determined, then welcome to the ensuing pages, where you will learn about pursuing your dream, beginning with the life and duties of an apprentice film editor in a major Hollywood studio.

CHAPTER 2

THE APPRENTICE FILM EDITOR

Those of you who are able to secure union jobs will begin working at a major studio or an independent firm. It will be helpful to understand the structure of such companies so that you will better comprehend who is responsible for whom and where you, the apprentice editor, and your fellow editors fit in. Figure 2–1 illustrates the position of editorial personnel in the studio organization from the perspective of episodic series television. (*Episodic series* are ½- and 1-hour weekly, prime-time programming.)

Since episodic television creates the most job opportunities for editing, I am going to concentrate on that area of filmmaking in the company but I will note some procedures that apply to other formats: miniseries, movies-for-television, and features.

It should be noted here that there may be some variations between Figure 2–1 and reality. Some companies have an executive producer as the administrative head or a supervising editor instead of a post-production supervisor; other companies have dispensed with their sound effects or music editing departments, subcontracting those functions to independent houses.

Let's review Figure 2–1. The administration hires the producers for each series and also hires the post-production supervisor. The producers hire the writers, directors, composers, and film editors and exercise approval of the crew. (In nonepisodic TV the director generally hires the editor, assistant director, art director, and director of photography and approves the crew and other personnel.)

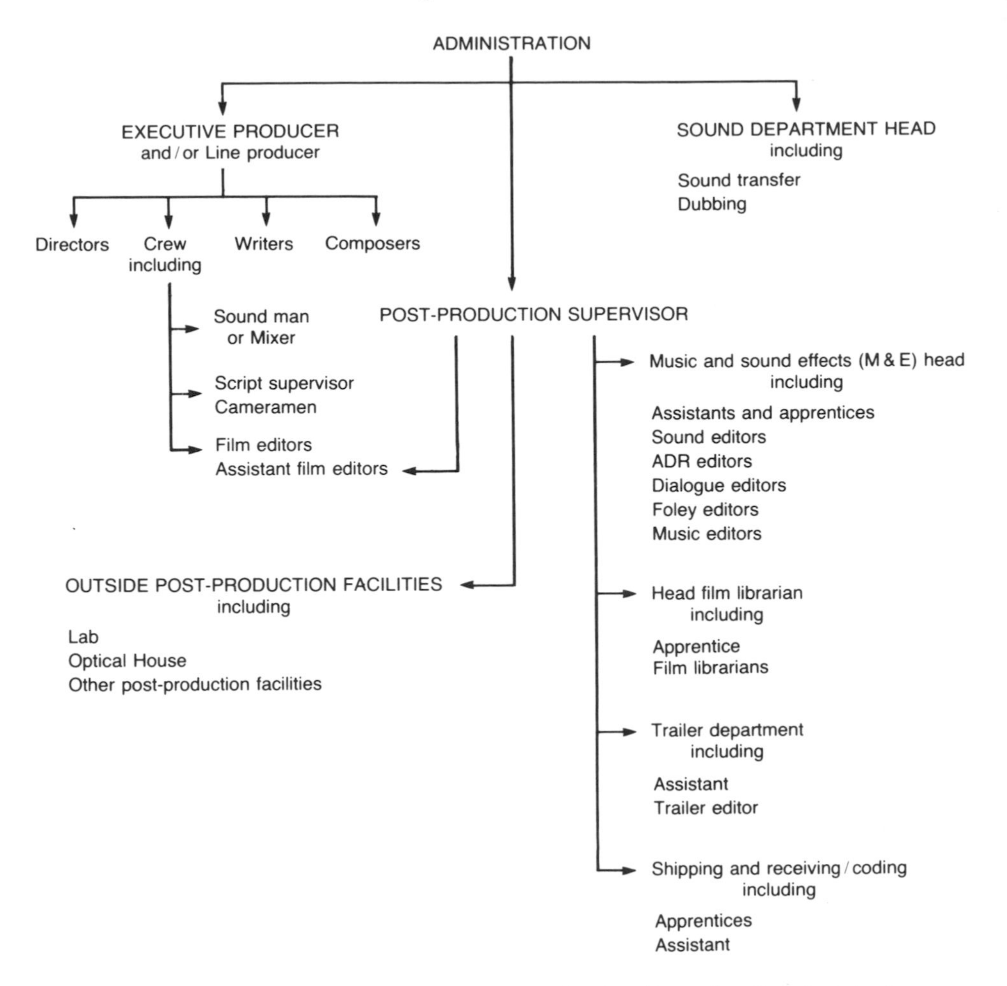

Figure 2–1. Major studio/company organization (for episodic television).

THE POST-PRODUCTION SUPERVISOR

Post-production supervisor is a misnomer for this far-reaching position. This supervisor is very involved with preproduction of a project, for instance, and is active during shooting as well. The administration hires this supervisor to oversee all television and theatrical projects sometimes assigning the two divisions to two supervisors. As will be discussed in more detail later in the text, the post-production supervisor also represents the administration with the lab, the optical companies, any other post-production facilities that might be employed, and, in certain matters, the producers of the various television projects.

The post-production supervisor is responsible for hiring personnel for his five departments: shipping and receiving (including coding), sound effects, music, film library, and trailer. As Figure 2–1 indicates, the supervisor is responsible for the hiring and assignment of all apprentices and assistants. He usually entrusts the hiring of librarians and music and sound editors to the department heads, however. The episodic series editors are chosen by the producers while directors of other projects generally select their picture editors. Should producers and directors have no specific person in mind they may ask the post-production supervisor for suggestions or simply delegate to him the hiring of a qualified person.

The post-production supervisor assigns assistants to the film editors. Should a staff assistant not be available or should the supervisor have no preference on the union availability list, he might then be receptive to a recommendation from the editor.

The primary concern and responsibility of the post-production supervisor is always to have the company's projects completed and ready for all scheduled television air dates or release dates for theatres. In order to achieve this, the supervisor will need personnel who are dependable and efficient, who can work together harmoniously as a team, and who will not create any problems for the producers, directors, or the supervisor himself. Except for those unavoidable situations involving politics or nepotism, the supervisor tries to be very selective when filling vacancies on the editorial staff. And now the supervisor has selected *you* as the new apprentice film editor.

BECOMING AN APPRENTICE

It is now your first day. You report at 9:00 A.M. to the post-production supervisor's office, where his secretary welcomes you and has you fill out your payroll card indicating your name, address, phone number, Social Security and income tax information, and any payroll deductions you wish for the studio credit union. Such automatic deductions are excellent forced savings for those rainy days of unemployment that ultimately come to all editorial personnel.*

Your are greeted by the post-production supervisor and you wonder: Where is he going to assign me? What are his choices? The supervisor can assign you to one of several different departments.

EDITING ROOM

You would be fortunate indeed to begin your apprenticeship in an editing room. It would be the best possible learning experience for a beginning editor.

As the apprentice, your principal job would be that of a *gofer,* transporting film and running various errands. If a coding machine were available, you would learn to code film. You would label trim boxes and learn to splice and rewind film. As the assistant editor and film editor gained confidence in your ability, they would begin to give

*At this point you might also agree to contribute to one or both of two worthy charities, the Motion Picture and Television Fund that is contributed to by members of the motion picture industry for the maintenance of the Motion Picture Country Home and Hospital for the industry's needy, infirm, and disabled or the Permanent Charities Committee of the Entertainment Industries (PCC)—a system founded by Samuel Goldwyn as one central organization to raise and distribute monies to a multitude of deserving charities in the Southern California area. Currently the PCC charities include over 400 charities (including the American Red Cross, 360 agencies, and 15 health partners serviced by United Way) and over 25 other independent charities.

you more responsible assignments. Under the assistant's supervision you will begin to learn everything pertaining to the *sync*ing up of *dailies* and the filing of *trims.*

If the editor is using a *flatbed* rather than a *moviola,* a major and time-consuming occupation of the apprentice would be the continual reconstitution of the so-called *Kem*® rolls.

But here's the bad news. It is very unlikely that the post-production supervisor will assign you to an editing room. It is more likely that he will assign one of his more experienced apprentices there and that you will be that person's replacement.

THE FILM LIBRARY

It is quite possible your first assignment will be as the film library apprentice. Usually, only one apprentice is needed to service the librarians so it would be helpful if the apprentice you are replacing can remain with you for a couple of days to break you in. Otherwise, you will have to rely on the busy librarians to teach you and answer your questions.

A film library's function is to supply stock footage required for any of the company's projects. If satisfactory material is not in the library, then the librarians try to obtain it from other film libraries. The apprentice is responsible for shipping and receiving, keeping records, and transporting film between the library and the editing and projection rooms.

Your other duties, under the direction of the librarians, include the maintenance of the film in the vaults and of the vault records, the repairing and splicing of film, and writing orders.

MUSIC AND EFFECTS

It is very possible that you will be assigned to the music and sound effects (M&E) pool where you will work with several other apprentices and assistants. While the roles of the music editor and the sound editor are clearly defined and rarely interchangeable, in your studio situation often less than half a dozen assistants and apprentices service 20 to 30 of the editors doing whatever sound effects and music chores that are required.

As one of these apprentices your chief duty is the transportation of film between your department and the sound transfer room, the coding room, and the dubbing stages. You are the messenger between the sound editors and the sound effects library. Under the supervision of an assistant, you will quickly learn to splice and build sound effects and music units.

Your responsibilities in and interaction with the editing rooms, film library, and music and effects pool, either as an apprentice or assistant, will be discussed more fully later.

SHIPPING AND RECEIVING

Most apprentices begin in shipping and receiving, which usually includes coding. In this case the post-production supervisor tells you to report to the assistant in charge, points you in the general direction, and wishes you good luck. You wend your way through a labyrinth of buildings and sound stages and, after several wrong turns, one of which leads you into a machine shop and another into a rest room, with the help of a few passers-by, you finally find your new home.

It is 9:30 A.M. The shipping and receiving department is already a flurry of activity. A lab truck has dropped off a delivery and assistants from various projects are grabbing their dailies, giving the assistant in charge—whom we shall also call department head or head assistant—barely time to check in the film properly. A couple of coding machines are running and telephones are ringing. One apprentice is loading film onto a golf cart while another has just returned on a bicycle from an errand. Another apprentice is at a bench splicing.

In half an hour things quiet down enough for you to be properly introduced to your co-workers and your training as an apprentice editor begins.

35MM FILM

Before I describe your duties and responsibilities as an apprentice in shipping and receiving, you first need to have some basic knowledge of film itself.

The major theatrical and television projects that we are concerned with in this text are edited on *35mm film.* The negative film has an emulsion coated onto a *cell* (cellulose) base by the stock manufacturer. It is placed in the magazine of the camera where it is exposed, frame by frame, during the shooting of a motion picture. When it is processed and developed after shooting, the negative is retained protectively by the lab or is sent to the negative cutter who has been assigned to the particular project. The negative film represents the original source material and must be very carefully handled and stored. When the negative is printed in contact with positive film, which has an emulsion coating upon which a positive image is produced, a *work print* is created (See Figure 2–2).

35mm photographic film contains 16 image frames per foot. Each frame has four *perfs* (perforations) or sprocket holes on both sides, and is referred to as a *four-sprocket frame* (*not* an eight-sprocket frame). Each frame is separated by a *frame line.* Sometimes the line is barely visible but there is usually a *frame marker,* in the form of a

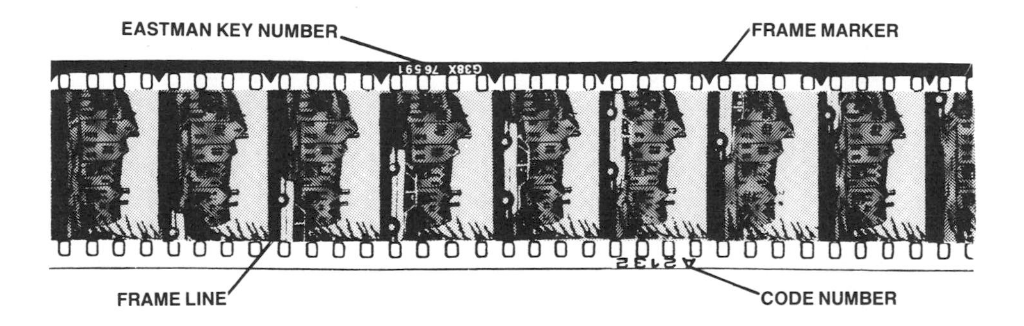

Figure 2–2. 35mm positive film.

small black triangle, either on each frame line or on every other one. These markers are also called *pips, pins,* or *tits.*

Standard film projection speed is 24 frames per second or 90 feet per minute. The film is run and edited with the *emulsion side topside and the cell side underneath.* The two sides are distinguishable because one side appears dull (the emulsion side) and the other side is shiny (the cell side). Should you have difficulty distinguishing the two sides, place the film between your lips. The side that clings to one lip will be the emulsion side. To date I have had no reports of any fatalities resulting from this experiment.

So the correct position for the film is *emulsion, dull side* or *face up* and *cell* or *shiny side down.* Placing the film in a horizontal position, as illustrated in Figure 2–2 and as you would on an editing bench, the film is going left to right with the beginning or *head* of the scene to your right and the end or *tail* of the scene on your left. If you turned the book to look at Figure 2–2 vertically, as you would in running the film *heads up* in a moviola, the left edge of the film has a black line through which can be seen *negative key numbers* imprinted from the negative. On the right there is a clear edge upon which your company imprints its own *code numbers.* (The term *edge numbers* should be used in reference to *key numbers* rather than *code numbers.*) Both key numbers and code numbers are printed in numerical succession every foot of film. Key numbers identify the film for all lab and optical work and negative cutting. Code numbers are used in all editing stages to identify the film and also to maintain synchronization between picture and sound. More will be said about that in the following chapter.

As I have previously noted, you will undoubtedly be working with color 35mm film. Black and white (B&W) film, which was the major film medium for over half a century, is only rarely used now although it is sometimes included in a sequence for nostalgic effect. A recent feature shot entirely in B&W was ''Raging Bull,'' released in 1980.

B&W film should not be confused with B&W reversals that are contact prints made not from any negative but directly from the color print. B&W reversals (also called *dupes* or *dirty dupes*) are used principally as reference copies of the edited work picture by music and sound effects and sometimes in ordering complicated opticals. Afterward, they are usually salvaged for use as *sound leader.* This will be discussed in more detail later.

You may have occasion to handle 16mm film, usually stock film or completed films reduced to 16mm. Outside the main prime-time and feature film production segment of the industry where you are working, 16mm is used to shoot documentaries, educational films, commercials, and a variety of projects where the major consideration is the lesser cost of 16mm film stock and processing as compared to 35mm. While the projection speed on both gauges is 24 frames per second (*fps*), 16mm has two-and-a-half times more frames per foot than 35mm—i.e., 40 frames versus 16 frames—and therefore runs at 36 feet per minute while the larger 35mm film runs at 90 feet per minute.

16mm has only one perforation at each frame line (see Figure 2–3). Being so much smaller than 35mm it is more difficult to work with and is often referred to as *spaghetti.* Film students, who generally work with 16mm in their cinema classes, find working with 35mm a pleasant relief. When 16mm stock is selected for use in a 35mm film, it must be optically blown up to 35mm.

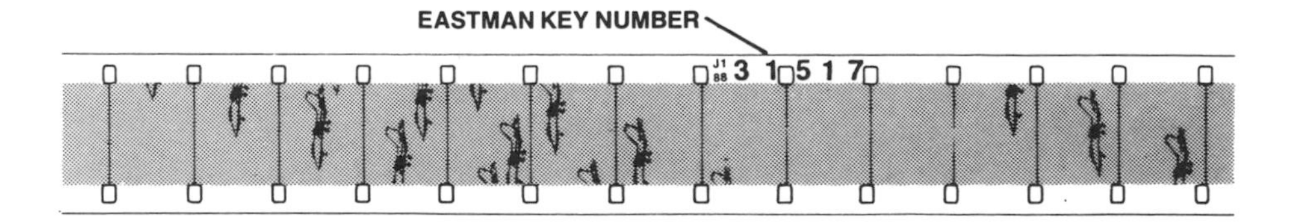

Figure 2–3. 16mm positive film.

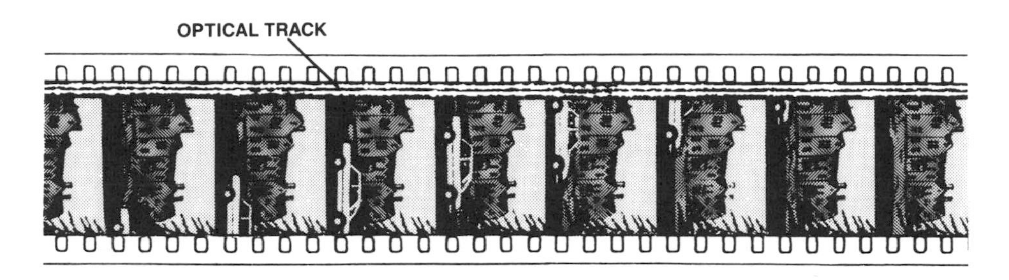

Figure 2–4. 35 mm compositive film.

After all editing and other post-production work has been accomplished, the completed picture, the film that is projected in your local theatre, is called a *composite print,* meaning that it contains both picture and sound track together on one film stock as illustrated in Figure 2–4. The wriggly lines on the left of the film comprise the *optical track* containing all the *dubbed* (or *mixed*) elements of dialogue, sound effects, and music in sync with the picture. This is the sole use of optical tracks. Instead, *magnetic sound tracks* are used in editing and post-production.

MAGNETIC SOUND TRACK

In the early days of talking pictures, film was edited with optical (or photographic) sound that was synchronized to the picture. The sound was photo-recorded on 35mm sound negative film and then transferred to 35mm positive stock with perforations corresponding to those on picture film. The track appeared on the left side, emulsion up, as in Figure 2–4, without the picture. It would show the pulsations of the highs and lows of the background sounds and dialogue so that an experienced assistant and editor, familiar with the script, could actually *read* those fluctuations and, by sight, know whether they were cutting before or after a certain sound or word. This is no longer possible since *magnetic (mag) tracks,* which cannot be read, have replaced optical tracks as work film.

Mag sound is recorded on an oxidized, emulsion-coated surface and also has perforations corresponding to 35mm picture. In most companies when you are editing with sound, the proper placement of the track is *emulsion* or *dull side up* and *cell* or *shiny side down* as with picture. This may also be referred to as being *mag up.* There are three types of mag track with which you should become familiar: *single-stripe, three-stripe,* and *full coat.*

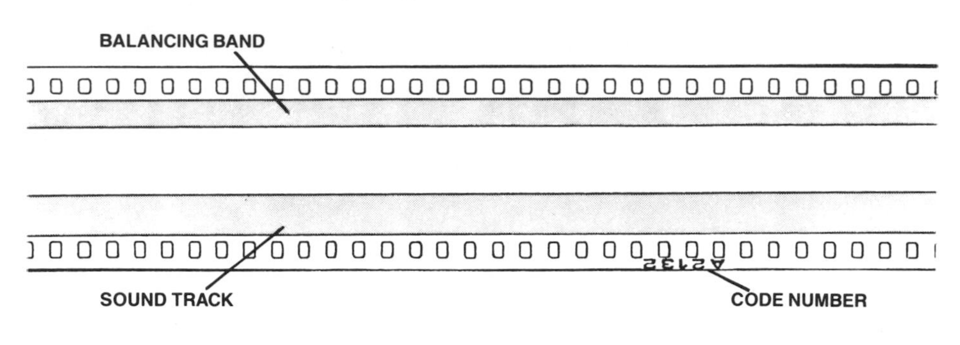

Figure 2–5. Single-stripe magnetic track.

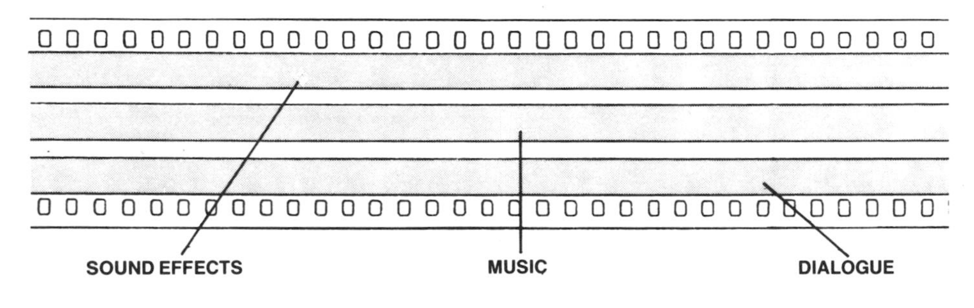

Figure 2–6. Three-stripe magnetic track.

Figure 2–7. Full-coat magnetic track.

Single-stripe mag with emulsion up (Figure 2–5) has the wide stripe with the sound on the right while on the left is a narrow *balancing band* that has no sound and only serves to maintain level contact with the sound head so that the best quality sound might be produced. This single-stripe mag is used as the *work track* (W/T) to be edited with the *work picture* (W/P).

Sound stripes are also called *channels, stems,* or *bands.* Three-stripe mag has three channels and was used by music and sound editors and in dubbing (see Chapter 16) where the right channel mag up was used for dialogue, the middle channel for music, and the left channel for sound effects. The use of three-stripe by major production companies, however, has been phased out, replaced with full coat mag, and included here only because it may still be used by someone somewhere.

Full coat mag is used as a master track in final dubbing and in sound effects libraries where it is treated in the same protective manner as picture negative. It has the emulsion across the entire width of the film, produces the best quality sound, and of course is the most expensive kind of track.

As an apprentice in shipping and receiving, your principal involvement with all these types of picture and track will be in transporting and coding them.

CHAPTER 3

SHIPPING AND RECEIVING AND CODING

The shipping and receiving department is usually located so that delivery trucks have easy parking and access. There must also be accommodations for transportation equipment such as golf carts, bicycles, and hand carts. This department should not be confused with the studio mail room (or messenger center) that handles all postal material and interoffice mailing. All film and orders or memos pertaining to film coming in or going out of the studio are the responsibility of shipping and receiving.

Records must be maintained of all outgoing film and of the receipt of incoming film. These records can be in the form of a duplicate copy of a delivery slip or of a list of items on a shipping invoice indicating time and date of receipt or departure. You must learn the company's system so that you, like other personnel, can properly check in and check out material.

There should be receptacles for incoming film, usually large wooden shelves with partitions labeled for each production and series. Each production, even an episode of a series, has its own production number. A delivery slip (see Figures 8–7A, 8–8A, and 8–9A) accompanying a carton of incoming film will indicate the production number so you will know where to place it.

Deliveries may be made at various times throughout the day and will consist of picture dailies and prints ordered from the labs, optical prints from optical houses, stock film from outside film libraries to be distributed by your film library, picture reels being returned to the editors by negative cutters, features being loaned to your studio, and stock film and features loaned by your company to other companies that are now being returned.

Assistant editors will probably pick up their film before you need to call them, but in any event do not let film remain on the delivery shelves for long periods without advising the proper recipients that their film is there, so that they will know it has arrived and will either come for the film themselves or instruct you where to deliver it.

The person in charge of scheduling projection room screenings will send your department a daily *schedule of screenings* for the entire studio or company. This person might be an assistant or secretary in the post-production supervisor's office or in the projection department. The schedule will list the time, projection room, and identification of film that is going to be viewed and will thus provide delivery destinations for some of the incoming film as well as for the film being worked on in the cutting rooms. Any changes in or additions to the schedule such as the running of dailies are called in and should be properly noted.

The shipping and receiving department head should make certain that you and your co-workers know your responsibilities regarding the schedule each day so that no one is kept waiting for film delivery. Film that is going to be screened should be delivered to the projection booth at least ten minutes before the designated time.

If you receive a feature from another company for a screening, it will probably be on 2,000-foot reels in two or three Goldberg cases (Figure 3–1) of varying sizes depending upon the length of the film and the number of reels. Open the cases and check the individual reels to verify that you have received the number of reels indicated on the front of the cases and that a reel of a different feature has not been mistakenly substituted. If you discover one of these errors and notify the person who has ordered the feature immediately, there might be time to correct the error before the screening. You are certainly not performing your job thoroughly if you wait until just before the

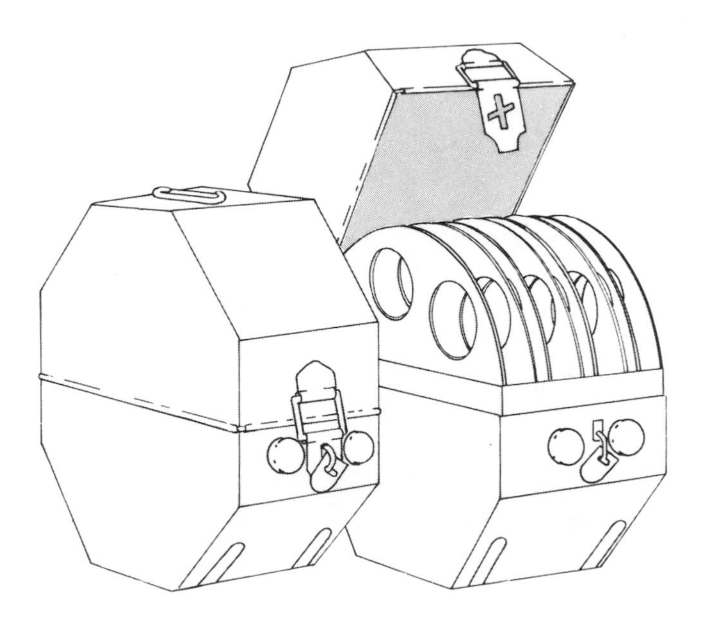

Figure 3–1. Goldberg cases with 200-foot reels.

screening to check the reels or do not check them at all so that the error is not discovered until the projectionist opens the cases.

Most of your deliveries to the projection booths will consist of picture and track, each on 1000-foot reels (Figure 3–2). These would be *dailies,* which are picture and track of the previous day's shooting *synchronized,* or *synced,* by the film's assistant editor; or the delivery could be the *cut picture,* or *work print,* the work picture and work track *cut,* or *edited,* by the film's editor.

Different companies, even different individuals in the same studio, may have different procedures for the delivery of these reels. In some cases, the assistant may want to deliver the reels herself and you will not be responsible; in others, either the assistant will bring the picture (meaning *both* picture and track) to shipping and receiving or you will have to pick it up in the cutting room when you are informed that it is ready. You therefore have to become familiar with each editor's procedures.

If you are in one of the larger studios, be prepared to schlepp film as far as several blocks using whatever conveyance might be available, the golf cart being the most desirable conveyance. You may then have to hand-carry the film up several flights of stairs to the projection booth. Do not try to lift or carry more film than you feel comfortable with. Better to make more than one trip than to injure your back.

If the editing room is in a bind with the screening and all the film is not ready in time to get it to the projection booth, you may have to *bicycle* the reels, taking the earlier ones first so that the screening can begin and then running back and forth for the remainder as the assistant editor completes each reel.

Shipping and receiving should have cartons, containers, or some sort of plastic covering for this film in the event of rain. Film should always be protected from water or any kind of liquid since it will blotch the film and can also make the film stick together thereby causing irreparable damage.

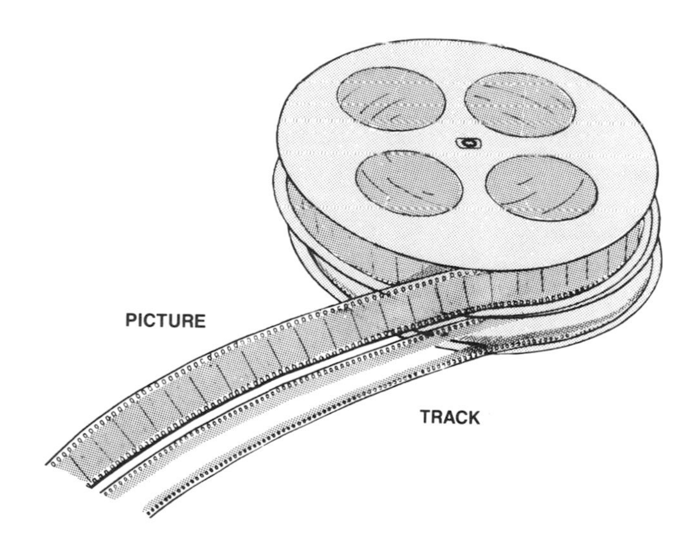

Figure 3–2. Picture and track on 1000-foot reels.

When you deliver these reels of picture and track to the projection booth, do *not* pile them up on the floor. Usually there is a *reel rack* on which to line them up left to right in consecutive order, beginning with Reel 1 Picture, Reel 1 Track, Reel 2 Picture, Reel 2 Track, and so forth. If there is no rack, line the reels up on a table or desk from top to bottom in consecutive order. After you have made your delivery, return to shipping and receiving because other errands or duties may be awaiting. Do not hang around the projection booth to see the film!

Either the running schedule or the assistant on the film will tell you what is to be done with the film after the screening. Perhaps there may be another running shortly afterward in which case either the film may simply be left in the projection booth or it may have to be taken to a different room. If the film will not be screened again, you will have to return the picture to the cutting room, or if it is dailies, schlepp it to the coding room for coding.

THREE TYPES OF CODING

In most studios and large companies the coding room is situated as close as possible to shipping and receiving so that it is managed by your department head, an assistant. The coding is usually assigned to you or your fellow apprentices on a rotating basis. This assignment may vary from a day to a week, and one or more apprentices may be assigned each *tour* depending on the amount of coding required. This does not mean that you may not be pulled away from the coding machines for an urgent delivery.

In the coding room there should be two clearly marked areas, either on shelves or reel racks, one marked *IN* for incoming reels to be coded and the other marked *OUT* for outgoing reels that have already been coded. If everyone maintains these separations correctly, much confusion and many embarrassing delays in coding will be avoided.

As I have previously noted, coding is imprinting numbers on the edges of both picture and track. There are two basic reasons for coding:

1. for synchronization of picture and sound and
2. for identification of the material.

This and many of the terms used in the description of coding in this chapter will be further explained later in relation to the assistant's duties.

There are basically three types of coding:

1. *standard,* or *American coding*
2. *numerical slate coding* and,
3. *Acmade® coding.*

Standard coding derives its name from the American coding machines that have been standard equipment in Hollywood for over 50 years, beginning with the single-reel system that codes one reel at a time (see Figure 3–3). Joining it in recent years

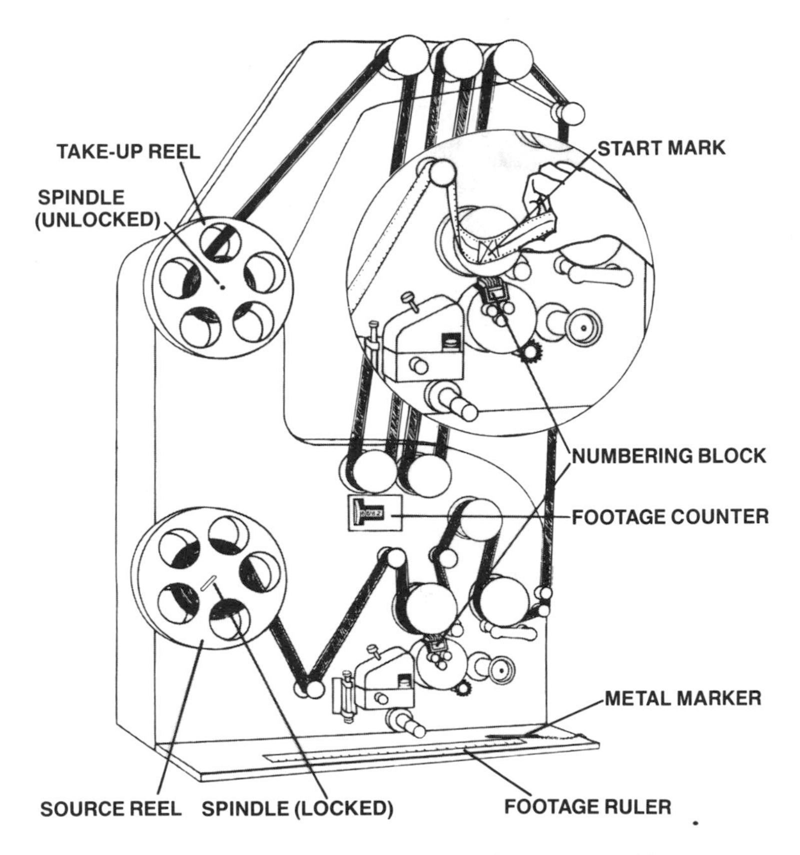

Figure 3–3. American single-reel coding machine.

has been the double-reel system on which picture and track can be coded simultaneously (see Figure 3–4).

Standard coding offers the choice of five or six characters, one or two letters (A through K except I), and four numerals. The assistant will indicate on the leaders of his daily reels the starting code numbers, the last three numerals customarily being 000. Therefore, a beginning daily code number would be, for example, A 1000, BF 5000, KA 0000, and so forth. CI 0000 could not be coded because "I" is not available. Neither could C 100 be coded because you must have *four* numerals.

All five or six digits are manually set on the *numbering block,* usually with a metal marker. As illustrated in Figures 3–3 and 3–4 the numbering block is a cyclical, rotating mechanism situated near the base of the coding machine. Once the desired beginning code number has been set in the printing area and the coding begins the single letter or two letters remain stationary while the four numerals will increase at a rate of one number for each foot of film.

These code numbers will appear on the right-hand clear edge of both picture and track and will occupy one frame (see Figure 2–2). There is no significance between any particular code number and the frame it happens to occupy other than, as previously

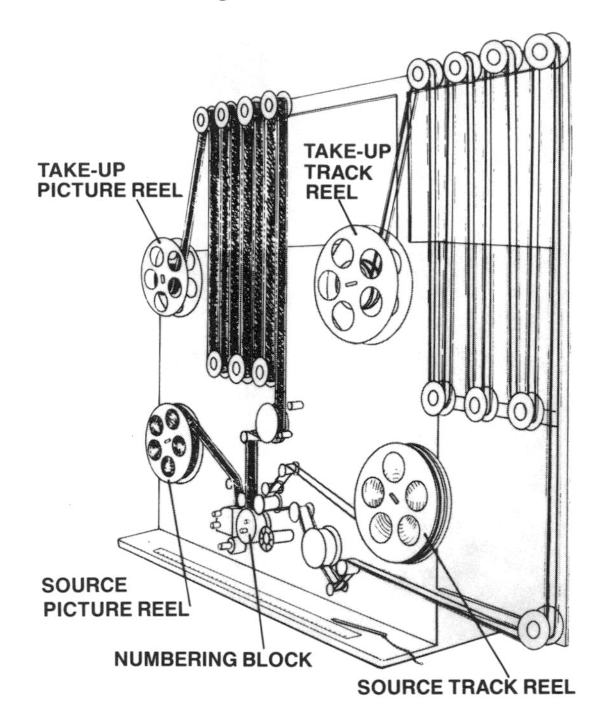

Figure 3–4. American double-system coding machine.

indicated, as a sync reference between picture and track and as identification of that entire take. (In Chapters 8–10 there are more detailed discussions of how scenes and takes are identified, the assistant's choice of code numbers, and her periodic use of their footages.)

While the so-called *standard* coding system I have just described is the one most commonly used in Hollywood, the above five- or six-digit numbering block can be replaced with any other numbers and letters that may be more desirable, if your company is willing to pay for such a special block. In addition, a standard machine with a different block can be adapted to the requirements of the *numerical slate system* that refers to the identification of each angle shot on the set by slate number instead of a scene number, with each take, or slate, *individually* coded to identify that slate number. (The assistant will have a cross-file identifying the scene number of each slate.)

This system uses, first, a numeral representing the take number and then adds one or two letters, each representing numerals, plus another numeral for the slate number. The letters, A through K except I, and the numerals they represent are as follows:

A	B	C	D	E	F	G	H	J	K
1	2	3	4	5	6	7	8	9	0

The above code system should be attached to the wall near the coding machines for easy reference by the apprentices. Again, the first number of the code is the take number, then the slate number, and finally the footage of each take beginning at zero.

Take	*Slate*	*Footage*
3	C5	000
1	BA1	025

Thus, in the first example above, the first code number, 3C5000, would translate to: slate 35, take 3, beginning at 0 footage. The second example would be read: slate 211, take 1, 25 feet into the take. Once the coding begins, the slate and take digits remain stationary throughout each take while the footage will increase with every foot of film.

Should the assistant decide to use the numerical coding block for standard coding, the apprentice need only leave the first digit blank to comply with any standard coding that might be requested. If, in the first example above, you omitted the take number 3, you would then have C5000. So, as you can see, using only the slate and footage characters, you can do standard coding.

It should be mentioned here that *black ink* is used for most American coding, particularly of dailies, simply because it is more permanent than any other color. *Yellow ink* is usually available for any additional coding such as music coding or preview coding.

While the American standard machine is a heavy, bulky piece of equipment and uses ink that prints the code numbers directly onto the film, the Acmade Codemaster is a light, easily portable single-reel machine that uses tape to emboss the numbers onto the film. Unlike the American machine that prints the numbers on the right side *edge* of the film within one frame, the Acmade code numbers are printed on the *left* side of the film *between the sprockets* extending onto a second frame. (It can print on the right side, but the numbers would have to be read from the reverse or cell side and would be bothersome and impractical for the editor and assistant.)

The Acmade code has seven or eight characters. It begins with three numerals and ends with four numerals. In between those, there can be a blank, a 1, or one of the following letters: A, B, C, D, E, M, W or P. Sample code numbers are: 002 E 3000, 316 1 7000 and 083 5624. Although restricted to a different choice of letters, the Acmade code can include standard coding but numerical slate coding would require ordering a special *British* block.

The tapes that emboss the code numbers come on cardboard cores in rolls about ¾ of an inch in diameter and a ½ inch in width. Since the machine only codes on part of the width of the tape, a roll of tape can be used for up to 9,000 feet of numbering by coding on both sides of the tape and then down the middle. White tape is generally used for coding dailies because it is so easily readable, but the assistant has his choice of seven other colors—blue, black, brown, yellow, orange, red, or green, all of which can be used for any additional coding.

CODING OPERATIONS

American Coding

Most of the coding that has to be done is on *dailies* or *rushes,* which is a British term. This is called *daily coding* and it has beginning code numbers such as C 5000 and BA 1000 as was discussed earlier in this chapter. The manner in which the assistant selects the code numbers is fully described in Chapter 6. American, or standard, coding refers to the American coding machine being used.

First, note the code number inscribed by the assistant editor at the head of the leader. Using a metal pencil-like pointer usually attached to the machine, the apprentice sets the correct code number—let's say C 5000—on the *numbering block.* Place the picture daily on the short shaft, or spindle, as shown in Figure 3–3, remembering to flip the movable end of the spindle thereby preventing the reel from slipping off during operation. (The double-reel machine in Figure 3–4 operates similarly with both picture and track.)

Pull down the leader so that you might place the start mark on the numbering block (see Figure 3–3). There should be notches or some sort of *guide mark* on the block indicating where the start mark should be placed so that the beginning code number, C 5000, would *presumably* be coded there. The code number may not actually be imprinted on the start mark but C 5001 should appear *one foot* away, C 5002 a foot further down, and so forth. (Some machines may be set so that C 5001 will be 1½ feet instead of 1 foot from the start, but then C 5002 will be 1 foot further down and coding will then continue normally with the code numbers 1 foot apart. This is acceptable for daily coding providing *both* picture and track are coded identically.)

On all coding systems, the film runs left to right through the block. To begin coding, before completing the threading up of the film on the take-up reel, flip the switch and let the film be coded with only a few beginning code numbers. Check to see whether you have the right code in the right place—either on the leader near the start mark or, if it is not readable because the leader has been reused several times, on the first couple of feet of the first scene past the leader. This precautionary method will help you avoid an error at the start.

There should be a one- or two-foot *frame ruler* or notches on the counter in front of the coding machine, so that by starting with your first readable code number and counting back to the start mark one or two feet at a time using the ruler, you can easily check that the first number is correct and on the correct frame.

Now, assured that you have set the correct code number, you can finish threading the film on the take-up reel. Before setting the machine into motion, however, be absolutely certain that you have flipped the locks at the ends of both spindles holding your reels. Should you fail to lock in a reel it will surely fall off during coding, tearing sections of film to shreds.

As your film is being coded, you should occasionally check the numbers to see whether they are clear and not over- or under-inked. You can do this without necessarily turning off the machine. You may have to stop the coding to dry off the block or add ink to the ink well. Should you have to leave the coding room, the coding can continue and the machine will stop automatically at the end of the reel or, when coding by numerical slate, at the end of a take.

If your assignment is to code *numerical slate* dailies, before you put any reel on the coding machine you will have to wind down every reel of picture and track (see Figures 4–1 and 4–2 and related instruction on the use of the *rewind* and *synchronizer*). Let us say there are 10 or 15 takes on a reel. The assistant editor marks start marks and slate/take identifications at the head of each of the takes. At the end of each take she will draw parallel lines down the centers of the film using white grease pencil on the picture and red marker on the sound track. These lines alert the apprentice/coder to *punch* the edges of the film at the end of each take. Machines that are altered to accommodate numerical slate coding are activated by the punched indentations to stop automatically, enabling you to reset the numbering block for the next take. You should not fail to punch the end of a take; otherwise the machine will continue coding through the next take, which will then be miscoded.

As you can see, this type of coding requires more work on the part of the apprentice than does standard coding, and it takes much more time. In a large studio, the number of projects that might need this type of coding could keep apprentices busy all day with just coding. Warner Brothers, Disney, and various independent production companies sometimes use the numerical slate coding system.

The American coder is a rugged, heavy-duty machine and is very reliable when it is properly serviced and maintained. Some of the first machines manufactured by Kodak over 50 years ago are still in use. Servicing is minimal but should be done every day. The numbering block should be removed every evening and immersed overnight in a cleaning fluid. Also, as noted before, even during operation the coding should be observed to see if numbers are smudging because of overinking in which case the block should be wiped clear. Whichever machines you might have in your coding room, familiarize yourself thoroughly with the manufacturer's operating and maintenance instructions in addition to listening to the directions given you by your co-workers.

Acmade Coding (See Figure 3–5)

Setting the numbering block (or head) and checking your beginning code number should be a similar procedure on the Acmade to what you did on the American coder. However, almost all the other coding operations are quite different.

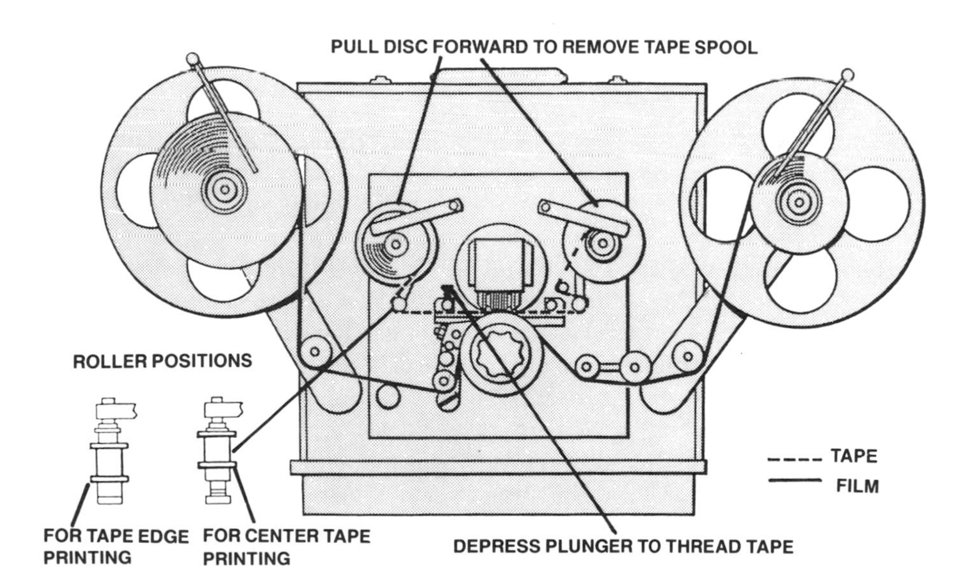

Figure 3–5. Acmade® coding machine.

The Acmade is not a heavy-duty machine like the American model. It is light and extremely delicate. It was originally designed to be used by a production company on a single project. In your studio, with many projects requiring continuous all-day coding, you should have several Acmade machines to spread out the burden. An over-burdened machine will develop problems and break down. Also, operation and maintenance instructions must be followed closely.

Before you begin coding allow the machine ¾ of an hour to one hour to heat up—its temperature should be no less than 155° centigrade and no more than 165°. There is a thermometer control on some machines and a built-in control on others.

While the American machines will code about 80 feet per minute and can manage 3,000 feet on any reel or core, the Acmade runs at a faster 100 feet per minute. However, you should not try to code more than 1,200 feet per reel as it will create too much stress on the machine. This will be an infrequent problem since most of your coding will be with 1,000-foot reels containing much less than 1,000 feet of film. If the film is on a core, it should be a three-inch, rather than a two-inch, core.

The numbering block need be removed and cleaned only after the coding of each 2,000,000 feet, but it should be oiled every other day, and the ball behind the block should be oiled at least weekly. You should examine the machine regularly and brush off dust particles that may have accumulated from the coding tape.

It is important that you thread your tape and film as instructed. Since the Acmade numbers are coded between the sprocket holes, the slightest variance may cause the numbers to move over a hole and become illegible.

When you are using the numerical slate system on the Acmade, the machine will stop only when the film runs out; punches do not stop Acmade machines. Instead of a punch the assistant editor must *cut* the film at the end of each take forcing the machine to stop. This relieves the apprentice of having to wind down the reels to punch each take before coding as you have to do when working with the American machine.

SPOT-CODING

Spot coding means coding the beginning number *exactly* on a particular start mark or indicated frame with no variance or exception. It is accomplished in various ways depending on your machine and on the procedure prescribed in your coding room. One popular method is to mark the film (lightly) one foot ahead of the beginning number indicated by the assistant, and then to begin *daily coding* there with one number less than the one requested. The desired number should then appear in the correct frame.

Let us imagine that an assistant has a reprint that is a duplicate of a take she had previously received. The reprint must be coded exactly as the first print and the assistant has marked it to be spot coded C 5738 on a particular frame or at a start mark on a temporary leader ahead of the slate. Using the above method you would set your numbering block for C5737 and begin daily-coding one foot *ahead* of that frame or start mark on which C 5738 should appear.

What if you are working on a machine which daily-codes the second number 1½ feet instead of 1 foot from the start mark, as described earlier? You would still set your numbering block at C 5737 but then you have to begin daily coding *1½ feet,* instead of one foot, ahead.

In addition to reprints, spot coding is usually required for:

1. prints ordered from outtakes,
2. wild tracks,
3. music coding, and
4. preview coding.

(These will be fully explained in later chapters.) It is also required for blocking out and recoding when correcting film that is coded out of sync.

CHAPTER 4

MORE APPRENTICE DUTIES

REWINDING

Coding sometimes entails rewinding the film. All coded material is normally left as it comes off the coder, *tails out,* and placed on the OUT shelf. A prime example of when rewinding is required is when the assistant informs you that the dailies you have coded are going to be screened afterward. Another example of a situation requiring rewinding, as described earlier, is before coding by numerical slate on an American machine. First you will have to run down the reels punching all the takes and then rewind the reels so that they might be coded from head to tail.

A shipping and receiving/coding department should have an editing bench equipped with such minimum necessities as *rewinds,* a *clamp,* a *synchronizer,* and a *butt (tape) splicer.* Of course, your room will have a supply of empty 1000-foot reels. Although it may not be necessary when transporting film or coding, before you start handling the film at the bench make it a habit to put a *white cotton glove* on your left hand. It will prevent you from scratching the film and may save you from a nasty cut from the film or from your equipment.

There are two rewinds on the bench, one on the left and one on the right. (The left rewind shown in Figure 4–1 has a swivel base that yours probably will not have but, instead, it would be like the rewind on the right.) A swivel base rewind is reserved for use with a moviola and will be discussed later.

The coded reels that you have been instructed to rewind are now tails out emulsion up. Place the reels on the shaft of the left-hand rewind so that the film will roll

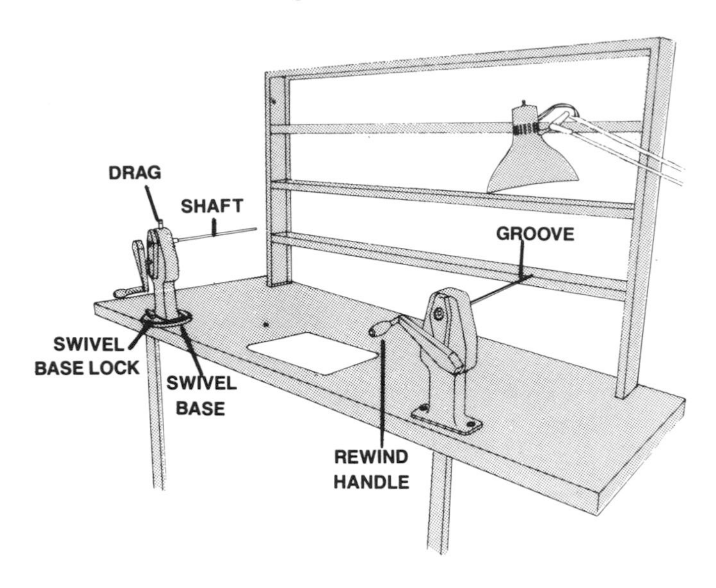

Figure 4–1. Rewind.

clockwise off the top of each reel. The length of the shaft dictates the number of reels you can rewind at one time. The average rewind equipment you are likely to find in your department and in editing rooms generally can satisfactorily rewind up to four reels at a time.

Thread up the film on an equal number of empty reels on your right rewind by inserting the ends of the film into the slots in the inside centers of the reels. Make certain that the foremost take-up reel is as close as possible to the rewind base by engaging the notch on its outside center to the small knob at the head of the shaft where it enters the base.

A *clamp,* also called a *spring clamp,* (Figure 4–2) has a *groove pin* that projects. By pressing a button on the side of the clamp, the pin will be recessed so that you can place it on the rear end of the right rewind shaft. Adjust the clamp so that when the button is released the pin will project into the groove of the shaft and you can guide the clamp forward against the take-up reels. As you do this, be careful you

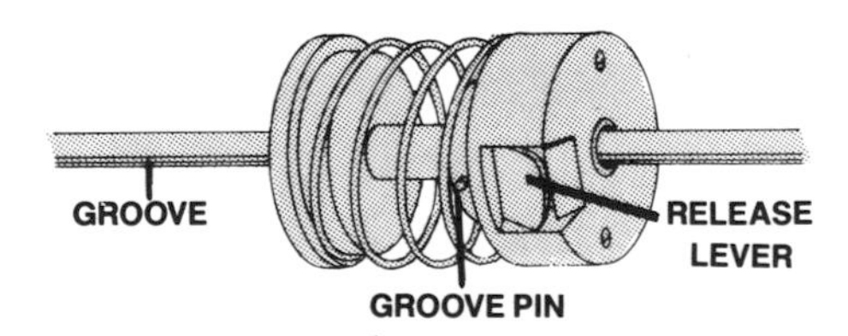

Figure 4–2. Clamp.

do not cut your fingers on the edges of the groove. The spring in the clamp will enable you to press it up against the reels with just enough pressure to make them rotate simultaneously. If too much pressure is exerted, the reels will bend and injure the film.

The handles with which you rewind can be disengaged from the shaft to facilitate its operation. When you are winding film from *left to right,* disengage the *left-hand* handle by pulling it *outward* from the rewind base, freeing the film so it can wind off easily. At the same time push the right-hand handle inward constantly to maintain its engagement with the shaft. As you keep turning the handle counterclockwise, the film will be winding up clockwise on the reels. Rest your left hand lightly on top of the take-off reels making sure the film rewinds smoothly and does not fly off the reels. As you rewind, try to keep the film fairly taut above the desk. If it is too tight, the film can tear and slack film can twist, catch on desk equipment, or drop to the floor and collect dirt. Should you have to reverse direction and wind the film right to left, simply reverse all the above instructions.

Do not try to rewind at maximum speed until you get the feel of it; you will become proficient very quickly. Most errors in rewinding that result in torn film usually happen to an experienced but overconfident worker through lack of concentration or pure carelessness.

Your department may be the proud proprietor of an *electric rewind.* You will not find it on an editing bench. Rather it will be on an abbreviated bench just wide enough for the normal distance between the two rewinds, and at a height that will require you to stand to operate it. At first glance it will appear to be an ordinary rewind, but on closer examination you will see a brake on the right rewind and a cord leading to a floor pedal from which runs an electric cord that is plugged into a wall socket. The slightest pressure of your right foot on the pedal will generate the right rewind to turn at a considerable speed, so exercise caution. Also it is safer to rewind only one reel at a time on this equipment.

When you have to stop the operation, release the pressure on the foot pedal while you press the brake down to stop the right rewind, and at the same time with your left hand stop the take-off reel from continuing to unwind. Since the left rewind is not driven electrically, this equipment is designed to be used only for left to right winding. The electric rewind is particularly useful in a coding room or a long-project editing room where a great deal of rewinding is usually required. Besides the advantage of speed the machine saves much wear and tear on the arm muscles.

Again, keep your left hand on the take-off reel to keep the film from slackening and to help slow it down should you have to use the brake on your right. Should you carelessly let a reel get away from you, this piece of equipment can cause a lot of damage to your film, so be careful.

CHECKING END SYNCS AND OUT-OF-SYNCS

The assistant editor inscribes *end syncs* on both picture and track at the tail end of the last take on each daily reel. The type of end sync symbol used depends on the individual

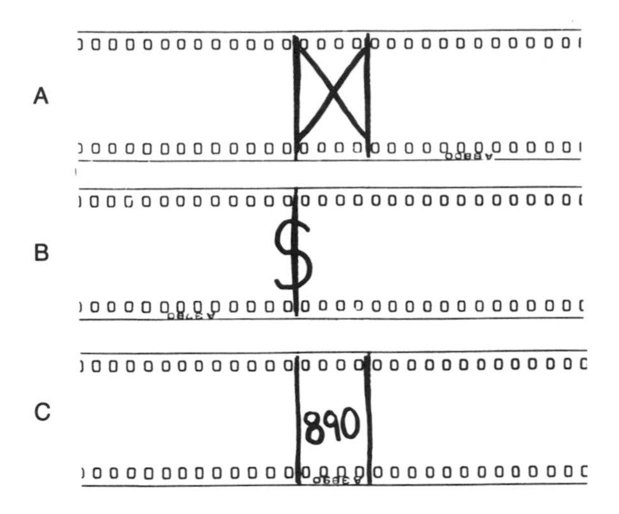

Figure 4–3. Three different kinds of end syncs (A, B, and C) commonly inscribed at the tails of the daily reels by assistants to verify that the reels have been coded correctly. While A and C occupy a full frame, B is on a frame line. C uses the last three digits of the code number that the assistant has already figured to be 890.

assistant's preference. The most popular symbols are illustrated in Figure 4–3. Figure 4–3c uses a three-figure footage number that should duplicate the last three digits of the code number in that frame. How did the assistant know the code number before the film had been coded? Good question and the solution is in Chapter 9 where the syncing of dailies is discussed.

End syncs on picture and track are exactly opposite one another. They provide a reference by which you can make certain the code numbers on both films match. Place one end sync over the other and then see if the nearest code numbers are aligned. If they are, the reels are in sync. If not, an error has been made (Figure 4–4).

If you have followed the procedure of checking the code numbers at the beginning of each reel, why check end syncs? How could the numbers be out of sync? They could be out of sync if the assistant has erred somewhere down in the reel or if the machine has *jumped.* If the assistant has used the three-digit end sync, you may immediately know whether the error is on picture or track because of the related code number. Regardless, you will have to determine exactly where the film went out of sync. You can only accomplish this by using a bench rewind and a synchronizer.

Ordinarily film that is heads up is placed on the left-hand rewind and wound left to right, heads to tails, across the bench as has been described. But remember that, having just coded the picture from head to tail, your picture and track are now tails out. Referring to the bench rewind in Figure 4–1, first place your *picture reel* onto the shaft of the *right-hand* rewind and place the track next to it at the rear. You will now be winding the film *right to left,* from tail to head, across the bench. Place both reels so

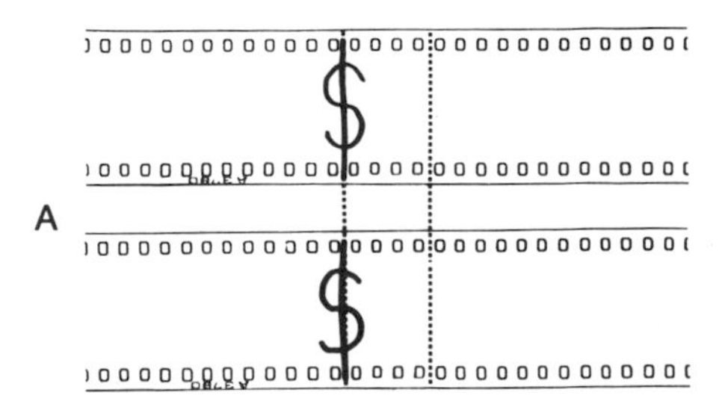

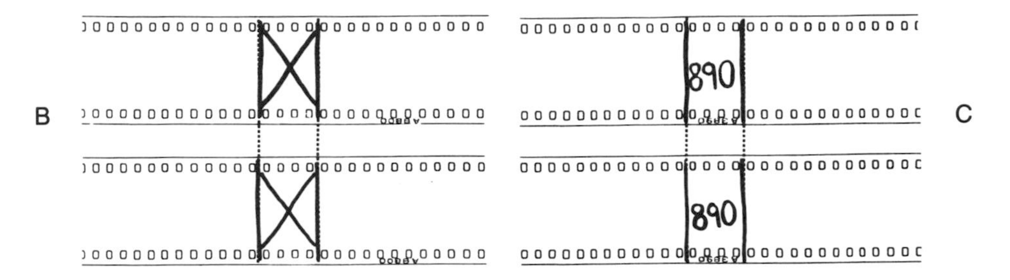

Figure 4–4. Matching code numbers by end syncs. This figure illustrates a hypothetical picture and track lined up at the three different end syncs. In each case the code number A 3890 of the picture and track is in direct alignment indicating that they have been coded in sync.

that the film will roll off counterclockwise from the *top* toward the center of the bench where your synchronizer is located.

Your *synchronizer* (Figure 4–5) is a *three-way* or *three-gang* machine providing positions for the picture in the front gang and the track in the second gang. (The use of the third gang is discussed in Chapter 9.) Note that the roller in front is partitioned into 16 frames per foot. Each frame is separated by a line and each gang has a small corresponding notch so that, as you place the picture in the first gang, you can match your picture frame line with either the line or the notch and the film will be in frame. Now place the track in the second gang, *lining up the end syncs exactly opposite each other.* The code numbers will *not* be opposite each other because, as has been established, they are out of sync.

You are now ready to attach the *tail leaders* (or ends) of your picture and track onto two take-up reels on your *left-hand* rewind. Place a clamp on the shaft against the two reels. In this hypothetical situation since you are winding right to left through the sync machine, you have to disengage the right-hand handle and keep the left-hand handle engaged while you are turning it clockwise. As you rewind, watch the code numbers in the center of your synchronizer; stop at the first point at which they correspond exactly and back up one foot. This should be where they went out of sync, because, as you remember, code numbers increase numerically every foot, and now it will be obvious whether it was the picture or track that was in error.

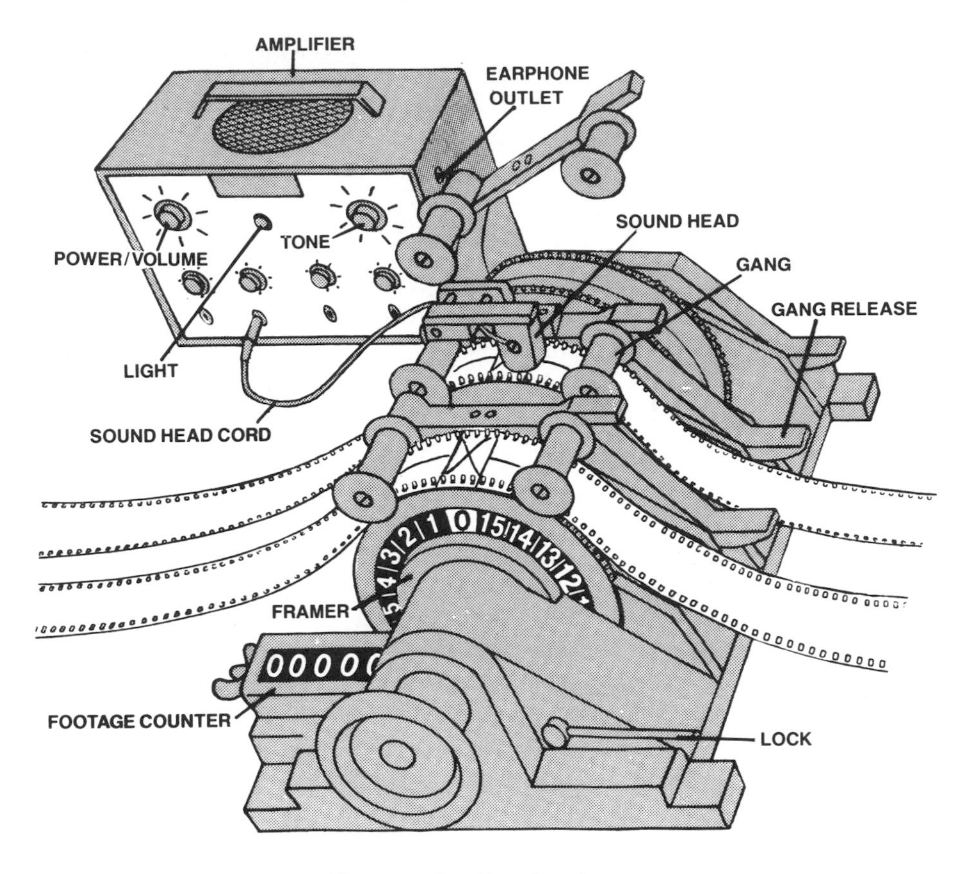

Figure 4–5. Synchronizer.

BLOCKING OUT

If the out-of-sync problem is the result of assistant error, you should contact the assistant and let him correct the sync before you adjust the coding. In any event you should always try to advise the assistant of any coding error before you make any correction. If it is a machine-caused error, have the machine serviced before using it again. Meanwhile use another coding machine to *block out* all the incorrect numbers by spot coding. Blocking out is simply selecting at random any code number and by spot coding positioning it so that it would print over or block out the first incorrect code number. By coding the blocking code numbers through to the end of the reel you will also be printing over the rest of the incorrect code numbers, rendering them illegible.

For example, let us say that the first incorrect number was after A 2735 and read A 2738. You decided to begin blocking it out with K 8243. It would look like Figure 4–6, which is certainly unreadable.

When you have completed blocking out to the end of the reel, rewind back again to where you located the error. From the last correct code number you can now

Figure 4–6. Blocking out.

spot code the rest of the reel correctly making certain that A 2736 is exactly one foot from A 2735. The codes on both reels as well as the end syncs should now be in sync.

Sometimes that erroneous number may be exactly on the correct sync frame or on an adjacent frame. Since a six-digit code number may overlap slightly into an adjacent frame, in either case it is impossible after blocking out to recode a completely clear number on the same frame. In that case you will be forced to block out *both* picture and track and then recode both, usually about eight frames away from the original coding so that there will be no confusion between the new code numbers and the blocked-out ones. This should be done, not from the spot where the mistake occurred, but rather from the slate or head of the particular take involved.

FILING AND SALVAGING

The shipping and receiving department is also generally responsible for the storage of film that your studio wishes to keep on hand. This film might consist of TV answer prints (see Chapter 16); of release prints of features; or of work prints or trims, lifts, and outs from these projects.

This film is usually kept in vaults or in some remote basement area with the location and identification of each movie or piece of film carefully catalogued so that any segment can be quickly retrieved if it is required. As an apprentice you will often be instructed to clean out an editing room and lug the film to the storage area. Film from episodic television series, since each episode has such a short editing life, will make up a large percentage of such material. By the time editing on three or four episodes has been completed, the film left over in trim cartons will be overcrowding the usually small 10 × 12-foot editing room, and the space must be cleared for more episodes.

Besides storing the film, you may be asked to salvage some of the reels that might be included with the material. Film that is to be saved, if it is on reels, is usually rolled onto cores using a *male flange* for takeup, and thus freeing valuable reels for more pertinent use.

The *flange* is a plastic or metal disk 11 inches in diameter with a center opening similar to the opening on a reel that permits it to be slid onto the shaft of a rewind. On the reverse side in the center is a cylindrical metal extension. On a male flange (Figure 4–7) this cylinder is grooved and has a retractable appendage on which a core may be attached.

A *core* (Figure 4–7) is a plastic cylinder, or center, that has an inner groove that permits it to attach itself to the appendage on the male flange that is on the rewind. On the outer side of the core there is a slit across the width similar to the slit within a reel. First, place the head end of the film into this opening to secure it and you can then roll the film on the core. Cores come in two-inch and three-inch dimensions with the two-inch core being the one most commonly used.

When you are winding onto a core, unless the flanges have a backing disk to enclose the core, you should limit your winding to one flange at a time. Backing disks are generally not available but *split reels* (Figure 4–8) are, and they make a more serviceable core-winding device. The backing disk of the metal split reel screws into the front disk enclosing the film you are winding on the core securely into position. With the help of a clamp, you can rewind up to four split reels at one time.

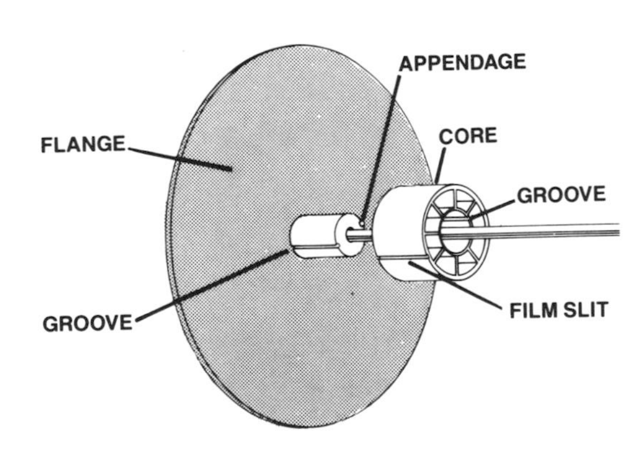

Figure 4–7. Male flange with 2-inch core.

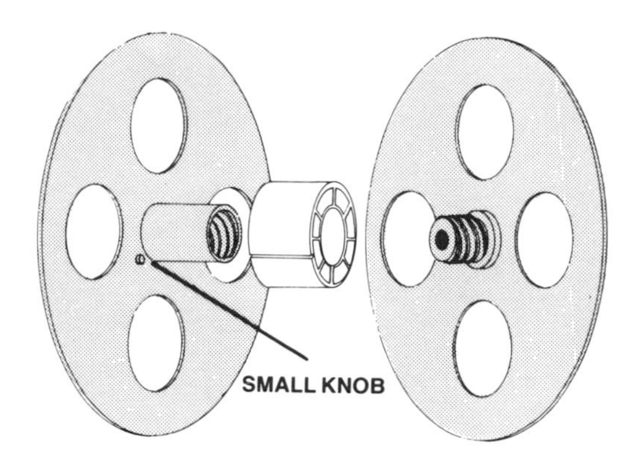

Figure 4–8. Split reel.

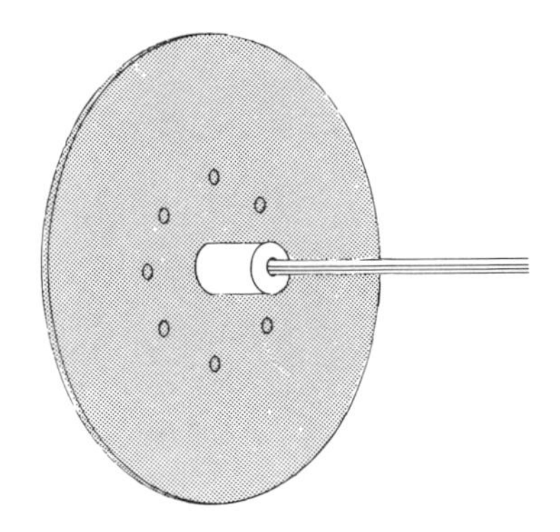

Figure 4–9. Female flange.

The post-production supervisor will eventually give his consent to discard certain film you have stored. Be sure you discard the correct film. Trim rolls (see Chapter 12) can simply be tossed into the trash container provided. But if you are required to save any rubber bands and paper clips that you may find in the trims, you will have to unwind the rolls into the container. In most major production companies discarded work picture, not trims, must be deposited in large black drums or barrels, separate from ordinary waste receptacles. These special waste bins are usually made accessible for pick up by film salvage companies that retrieve the silver from the positive prints. When salvaging reels or cores from waste film, you must use a *female flange* to unwind the film.

A female flange (Figure 4–9) is identified by a metal extension that is completely smooth so that film can be directly wound up on it without a core. Visualize that the reel or core you wish to salvage is on the left-hand rewind and you are winding the film left to right onto the flange on the right rewind. You would be turning the right handle counterclockwise. To remove the film from the flange turn the handle of the rewind in the opposite direction and gently but firmly remove the film from the flange. If you fail to follow this procedure, the film will simply tighten onto the flange.

Female flanges are not usually available but a small plastic *adaptor* (Figure 4–10) helps solve the problem. Inside this plastic cylinder is a ridge that fits into the grooved center of the male flange. Since this adaptor has a smooth surface, when it is attached to a male flange the latter is literally transformed into a female flange. Employ the same procedure as described above to remove the film from the adaptor—turn the rewind *clockwise.*

Sound effects units (reels) consisting of *black and white* (B&W) *reversals, work tracks* (W/T), *pre-dub tracks, color work prints,* or *release prints* discarded from completed projects will provide a large quantity of potentially salvageable material. The reversals, which are duplicate B&W prints made directly from *positive,* or work, *prints,* can be used as track leader, providing minimal time spent in splicing to make it useable; any more than half a dozen splices per 1,000 feet, for example, might be considered by your company to be excessive and not time well spent.

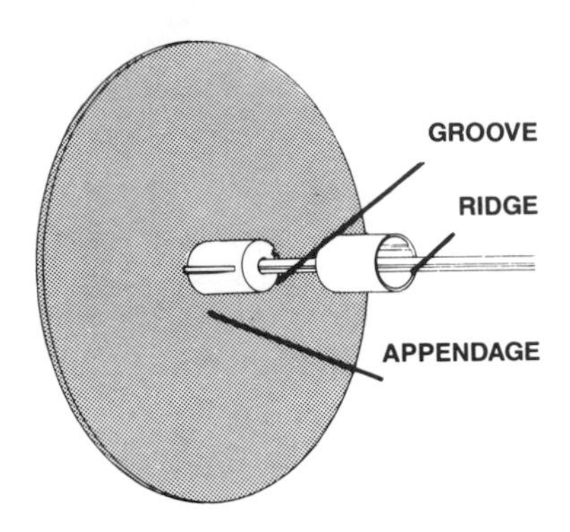

Figure 4–10. Adaptor on a male flange.

Each company has its own criteria about what should be salvaged. Generally speaking most major companies will not bother saving tracks that have more than one splice per 300 feet because of the danger of *pops,* disturbing sounds caused by defective splices. Smaller companies, for economic reasons, may not be that restrictive.

Such salvaged tracks are sent to your sound department to be *degaussed* (electrically erased) and will then be reusable, usually for certain sound effects. All saved material is rolled onto cores, using a male flange for takeup thus freeing valuable reels for more pertinent use.

SPLICING

Until the early 1960s, splicing work prints and work tracks was accomplished on *hot splicers* (Figure 4–11). In a large company or studio a number of these huge, bulky machines would sometimes be placed in one large room where apprentices would spend most of their day splicing. Although they have been replaced in the cutting rooms by *butt splicers,* an apprentice or assistant who may be assigned to a film library where hot splicers are still being used for splicing negatives or interpositives should be informed about them.

The apprentice or assistant splicing the film sits at the hot splicer, which is attached to a metal bench, and operates it using foot pedals. As is the case on an editing bench, rewinds are located on either side of the splicing machine's base and film is wound left to right as you splice. The picture is cut with scissors in the middle of a frame. In order to accommodate the requirements of the hot splicer two sprockets, or half a frame, from each piece of film are lost to produce a splice on the *frame line.* (Remember this point as it is very important and will be referred to again when editing is discussed in relation to negative cutting in Chapter 13.)

When hot splicers were used in editing to indicate the splices he wanted, the editor attached the end of one cut to the beginning of the next cut with a paper clip so that a reel showing a lot of paper clips was referred to as having *a lot of iron,* meaning you would have a lot of splicing to do. When the hot splicer operator came to a

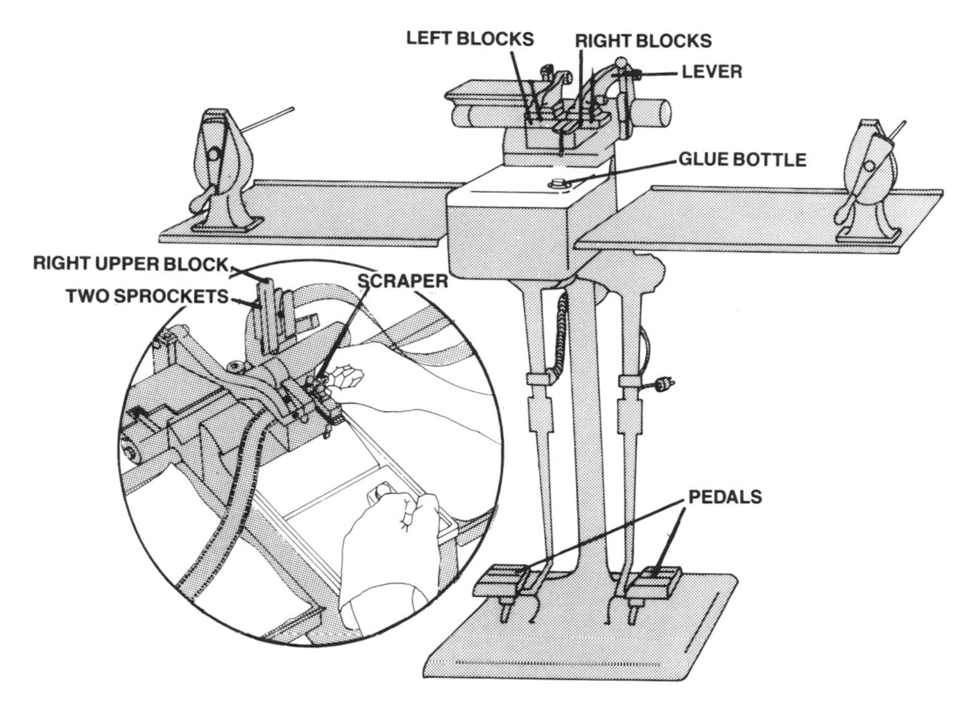

Figure 4–11. Hot splicer.

cut, he would remove the clip and, using the foot pedals, would raise the two upper splicing blocks. The end of the film on the right would be placed on the bottom side of the right block with two sprockets overlapping the center. The right upper block would then be brought down upon the film. Using a lever corresponding to the right-hand blocks, the upper and lower blocks were clamped together, securing the film between them. At the same time, by operating the right foot pedal, the clamped blocks on the right would be slightly raised. This operation was repeated for the identical mechanism on the left-hand side of the machine, except that the left-hand blocks were not elevated, but two sprockets of the film on the left also overlapped.

The apprentice then removed the emulsion coating at the frame line of the film on the left with a scraper; he also scraped the back or cell side of the elevated film on the right lightly. After brushing film cement onto the scraped emulsion, the operator lowered the right block. On contact the two overlapping sprockets on each piece of film would be cut off, and simultaneously the two pieces would be glued together by the heat generated by the electrical element in the machine. Residual glue would occasionally have to be cleaned off the splicer with acetone, a highly flammable solvent that you had to avoid getting on any film.

Although hot splicers are still essential to negative cutters, labs, and optical houses, as well as to film libraries, in editing they have been replaced by *tape,* or *butt, splicers* (Figure 4–12). They are small enough to be carried in one hand and portable

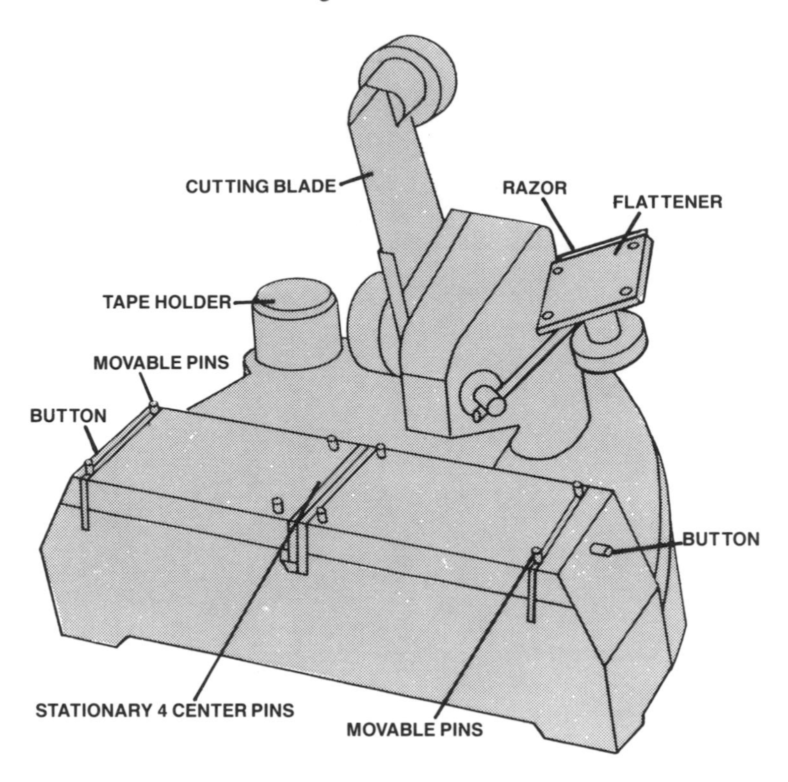

Figure 4–12. Butt splicer.

enough to be placed wherever one wishes on an editing bench. As I have already noted, you will have to do some splicing when you are salvaging film. Unfortunately, there will be occasions when film will break in coding, and you will have to do some repair splicing. When the more complicated hot splicers were used, apprentices did most of the splicing. Since the current tape splicing is much simpler and faster, the apprentice in shipping and receiving generally only splices for an assistant who needs some emergency help.

The editor does not cut the film across the middle of a frame with scissors as done with a hot splicer. Instead an upright blade, located at the center of the splicer, lands between the four pins on the base when it is pressed downward, and cuts the picture directly on the frame line. The four pins form an imaginary frame with two vertically aligned pins on the right of the center and two similar pins on the left. As with the hot splicer, the two pieces of film to be joined are held together with a paper clip.

Place the picture film on your right up *against* the two center pins on the left. You are now left with only the latter pins on which to place the end sprockets of the film on your left. The pins on both extreme ends of the splicer are self-moving or are made movable by pressing buttons on the sides of the splicer and will assist you in positioning the film.

Slightly to the right of the cutting blade is an upright blade lever or *flattener* (see Figure 4–12). On the left side of this lever a very sharp single-edged razor blade is attached. Be very careful with that razor; it can cut you badly. Bring down the flattener on the film to make certain the film is flat against the base. The razor on the lever is slightly raised and will not cut the film unless you purposefully pull the film up against it, and you would never do that. Now you are ready to splice.

You will have two rolls of 35mm splicing tape that are sprocketed exactly like your film. The *clear* tape is for *picture* and the *white* tape is for *track*. Since you are splicing picture, pick up the clear tape in your left hand, which you should position at least one-and-a-half inches from the center of the splicer. On picture, your objective is to make a *two-sprocket* splice on *both* the emulsion and cell sides of the film (see Figure 4–13.)

Place the *second* sprockets of the clear tape over the left center pins, forcing the first sprockets of the tape to overlap the splice. Bring the flattener down, adhering the tape to the film and at the same time tearing the tape toward you against the razor edge. Make the same two-sprocket splice on the reverse side of the film.

Unlike picture, track need be spliced only on one side, the cell side. The two ends to be spliced should be placed exactly in the center, with the second sprocket of each end on the center pins. The end sprocket of the white track tape should be on the right center pins making a *four-sprocket* splice (see Figure 4–14). Then follow the same procedure as for picture splicing, using the flattener and cutting the tape.

Once you have cut the tape, picture or track, always rub the tape with your finger to eliminate any possible bubbles, then gingerly lift the film off the splicer with both hands and place it on the desk. Finally, rub the tape again to make certain it is sticking firmly.

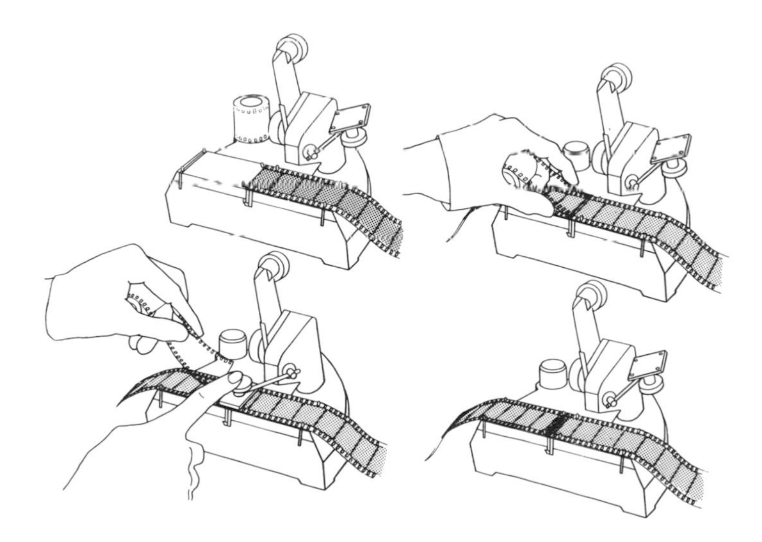

Figure 4–13. Splicing picture.

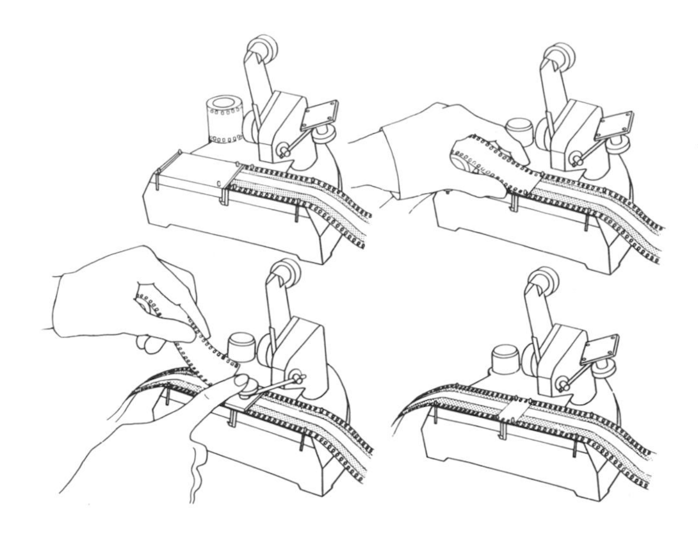

Figure 4–14. Splicing track.

Back in the 1960s, when tape replaced cement, picture splices required two frames so that both edges of the tape would be on a frame line that was not visible when screening. Some years later experiments with only one sprocket of tape on either side of the splice showed that the tape edges were still not visible and the two-sprocket splice became the norm. It was faster and, more important, saved quite a bit of picture tape. Some editors, however, particularly those working on long projects involving many screenings, still prefer the two-frame splice. Do not be afraid to ask what is required in a given situation.

CLEANING FILM

An assistant may enlist you to clean his picture reels before an important screening. This is primarily to remove dust that has accumulated during editing. A velvet cloth was used for this purpose in the past, but now this method has been replaced by Webril® wipes, which come in large rolls like paper towels, each roll containing 100 disposable cotton wipes.

As you wind the film left to right with your right hand, hold the wipe in your left hand, applying it lightly to both sides of the film and removing all dust. Do not press too hard or you may remove or smear editorial markings. In addition do not wind fast, and pause periodically to shake the dust from your wipe. When the wipe becomes overly dirty, discard it and use a new one.

Any recommended film cleaner such as a carbon tetrachloride solvent can be purchased from a film supply store and can be used on the picture to help eliminate a particularly bad mark or to assist with the dusting, but only a few drops on the wipe are needed at a time. Be very careful, however, to avoid dropping any cleaning fluid on sound; cleaners will destroy the magnetic track. A very dirty picture with dust ingrained in it should be sent to the lab for cleaning.

SUMMARY

Well, there you have it—all the tasks you may be required to perform, how to do them, and why they need to be done. As an apprentice in shipping and receiving/coding, you may be introduced to *all* of them in your first few weeks. Do not be frightened, and do not despair. They are neither as complicated nor as difficult as they may seem. If you remember the information given in this textbook, you will be surprised at how quickly and easily you will be able to perform your assignments—and with a minimum of instruction.

You have previously been warned and will continue to be reminded that some procedures vary from studio to studio and company to company. With regard to apprenticing, for example, not all companies locate coding in or near the shipping and receiving department. For example, at present Paramount has its coding machines near the editing rooms and, if the assistants do not have the time to code, a roving apprentice may be assigned to that job. A couple of years ago, before they transferred from film to videotape editing on their television series, Lorimar had quite a different, and laudable, procedure for training its apprentices: It periodically rotated them from one department to another so that they experienced everything. From a department's viewpoint this may be occasionally disruptive, but a completely trained apprentice crew provides obvious benefits to all the departments as well as to the post-production supervisor and to the apprentices themselves.

Also, while the instructions provided here refer to common practice, they are definitely not set in cement. You may find yourself working in a company where you are given instructions in some area that differ considerably from these. You must obviously follow the directions given you by your employer in such cases. Be open minded. Be prepared to adapt and adjust. This is true for any level of editing from apprentice to full editor.

More advice for you as you begin your first assignment: Be conscientious about your work, reliable, attentive, and cooperative and try to be reasonably pleasant and cheerful. Maintain these qualities whether you are on the same job for two months or two years. These may be old bromides that you do not think need even be mentioned, but you would be amazed at how many beginning apprentices, despite their delight at getting their first assignment, bring with them some dubious habits.

An example is the apprentice who is constantly late for work, but "only ten minutes. What's the big deal?" It *is* a big deal because in those few minutes a co-worker may have to interrupt his own work to cover for the late person, and this makes no one very happy, not the co-worker, the department head, or the post-production supervisor. Make it a habit to be *a little early* rather than on time.

Another example is the apprentice who, in addition to having studied this and other editing books, has taken editing classes and may even have had nonunion editing experience. This apprentice may act like a know-it-all. As a result this apprentice is not as attentive as he should be and misunderstands instructions or acts as though his talents and abilities are being wasted in inferior jobs. This apprentice may feel that becoming an editor is simply a matter of killing time.

Do not be one of those kinds of apprentice. Realize that promotions are not automatically handed to you—they must be earned. You are indeed fortunate to be a member of a marvelous industry and involved in a fascinating craft. You have the opportunity to make your dreams into realities.

CHAPTER 5

YOUR GUILD AND PREPARING FOR PROMOTION

YOUR GUILD

Once you have completed 30 working days in a 365-day period for a single signator employer or 90 days in the same period for more than one employer, you and your employer can file the necessary papers at that time with the Contract Services Administration Trust Fund, as was explained in Chapter 1, for your placement on the roster. Upon appropriate verification of your eligibility, your name will then be placed on the *Industry Experience Roster* and you must then apply for membership in the Motion Picture and Videotape Editors Guild, Local 776, IATSE, hereinafter referred to as simply the Guild or as 776.

As was indicated in Chapter 1, there are other unions that provide editorial personnel but 776, which celebrated its fiftieth anniversary in November, 1987, has about 2700 active dues-paying members and is the largest single-craft union in the industry. It is unequivocally predominant in the industry from an editorial perspective.

The primary objective of the Guild is to act as a collective bargaining agent for you and all your fellow members with the goal of maintaining fair and reasonable economic and working conditions.

The operation of the Guild is supervised by officers and a board of directors elected from the active membership by the members. The Guild office is maintained and its daily business conducted by an executive director, other representatives, and office personnel—all hired by the board of directors. The executive director also serves as business agent and is guided by established policies in representing the Guild and answerable to the board of directors.

The officers of the Guild consist of a president, a vice president, a secretary, a treasurer, and a sergeant at arms, all of whom are elected for two-year terms. To be an officer one must have been an active Guild member in good standing for at least three years.

In addition to the officers, the board of directors, representing various classifications of the Guild, numbers 26: 7 film editors, 6 assistant editors, 3 music editors, 3 sound editors, 2 TV station editors, 1 animation editor, 1 librarian, 1 apprentice, 1 videotape editor, and 1 San Francisco area representative. The directors are elected to staggered three-year terms, and they must have been Guild members for two years.

There is an annual membership meeting at which nominations are made for any board vacancies and/or for officers. Sometimes a special meeting is called. Besides regular board meetings the directors attend meetings of various committees on which they serve.

A member who is willing to devote effort and spend valuable spare time as a director or an officer is truly being altruistic. The only personal reward such a member can expect is the opportunity to be involved in the innermost affairs of 776 and to be able to meet many other members she otherwise might not get to know. The board member might also have the satisfaction of having tried to accomplish certain things for the betterment of the membership.

As a new Guild member, you still have a couple of years before you can run for the board. But you can become involved by attending Guild meetings, volunteering your services for a committee, or serving on the *Newsletter* staff. You can also take advantage of some of the activities offered such as screenings of recent theatrical films and informational seminars or of enrolling in periodic film training courses that are offered to Guild members at no cost. A videotape editing course is also available to members at a reduced fee. These programs will be not only enlightening and enjoyable but will also enable you to expand your career opportunities.

Now, at last, we come to what was probably your first question: What about the economics? How much is an apprentice paid, and what are his payroll deductions? Does it cost anything to join the Guild, and what are the dues? Are there any other benefits? These are all pertinent questions deserving considered answers.

You must first be aware that new bargaining agreements affecting wages and other benefits are made every three years and these result in changes every year. Therefore any figures mentioned here will be dated by the time you read them. To play it safe and avoid misrepresentation, I will give approximate figures.

As an apprentice, you will begin at a base salary of over $650. for a 40-hour week and will earn overtime should you work more than 40 hours. Payroll deductions include standard federal and state withholding taxes, state disability, Social Security, and Motion Picture Industry Pension. Voluntary deductions may be made to your studio's federal credit union, and charitable contributions of one percent or more of your salary (see Chapter 2). If you prefer, instead of payroll deductions you can make direct contributions to worthy charities.

In addition to your weekly salary most studios will give you earned vacation and holiday pay after March fifteenth of the year following your employment date. Some companies will prorate this pay and add it to your weekly check while others prefer to give it to their employees upon conclusion of their employment. Your employer contributes to your pension and to Motion Picture Health and Welfare, an insurance and health care program that provides you and your dependents with coverage for medical services, hospital, prescriptions, and vision and dental care. It is presently required that an employee work a minimum of 300 hours within a six-month period to be fully eligible for Health and Welfare (H&W) coverage the following six months. Also, should you work for one company for more than a year and then be laid off, you may qualify for severance pay.

When you submit your application for membership in 776 as an apprentice, you must include an entrance fee that has been $1000., so be prepared. In addition, your quarterly dues amount to approximately eight percent of your weekly salary. Since the apprentice's base salary in 1987–88 was $668. weekly, your dues would have been $54. every three months. The fee and dues are increased according to your classification.

Your Guild maintains an availability roster in each classification for employers seeking editorial help. Should you be laid off, it behooves you to notify the Guild immediately and to reaffirm your availability each month by postcard so that employers will know you are interested in employment. Conversely it is to your benefit to notify the Guild immediately when and where you are employed so that prospective employers do not make futile calls concerning present employment but can contact you regarding future assignments.

It is important that you put aside a reasonable amount of your earnings for the rainy days of unemployment. Your studio credit union provides you with an ideal opportunity for forced savings. For most employees in the industry there are always some rainy days every year. Television provides the most editorial employment and its busy season runs from about August to January, when production slackens for a month or two. It increases somewhat between February and April when television pilots for the new season are being shot. From then until about July, there is generally massive unemployment, referred to as a *hiatus*. (Actors say they are *between engagements.*)

Post-production supervisors do not enjoy laying off any members of their staff. They will try to keep you employed if you have proven to be a worthy employee so long as there is sufficient work. However, you will probably be the first apprentice your supervisor will have to lay off since, as a new apprentice, you are low person on the totem pole. And, although your supervisor will try to bring back everyone as soon as possible, you may be the last to return, or there may not be enough production to warrant calling you back at all.

Should you be laid off without any assurance that you will be returning, you should call the Guild immediately and have your name placed on the apprentice availability list. Your Guild's responsibility is to send a copy of the list to any signator employer upon request, not to find you a job. You will have to do that yourself.

Before your job ends advise any contacts you have that you will be available. It has also proven to be easier to learn about potential openings and possibly to find a job while still working at a studio than when you are unemployed. Do not be reluctant about requesting help from your co-workers or even your supervisor. They may make some calls on your behalf and find you some leads. I recall that when I was an apprentice in a film library and was layed off, the head librarian found an apprentice position for me at another studio until I was able to return to the library.

If you are able, it is wise to have some occupation outside the industry to fall back on should you have a long unemployment. It is difficult to find employers in areas other than motion pictures willing to train and hire you if they know you will leave them as soon as you find a film job. However, you do have some options if you do not mind work that may seem to you less attractive than you would like with a salary considerably lower than your studio wages.

Some restaurants may hire you as a waiter or waitress, or, if you can type or have some computer experience, you might obtain temporary clerical employment. Many actors and actresses have found the selling of real estate to be a lucrative second profession *between engagements* and some of them finally made it their primary source of income. Several editors with construction experience have made successful second careers of buying neglected homes, remodeling them, and selling them for respectable profits. Would you like to be a landlord or landlady? Find an apartment or condominium complex that will give you a rent-free apartment or condo in return for assuming the proprietary responsibilities. Another option is to have a relative or friend who manages or owns some sort of business and who is willing to employ you whenever you require a temporary job. You will be sustained through any period of unemployment if you truly possess the three P's—*passion, patience* and *perseverance.*

Until the 1960s there was a certain amount of permanency in Hollywood employment. Many studio people remained at one studio throughout their entire careers. They were a *family* and when it appeared that some members might have to be laid off, an effort would sometimes be made by the administration to create work to minimize the layoffs. It was a wonderfully protective and secure world but it also created a group of employees who, when they had to leave, discovered that few knew them at other companies and that therefore jobs were difficult to get.

It is a much different world today—mostly a transient one—in which personnel may come into a studio to do one project and then move on to another project at another studio. They may be doing a studio project or they could be working for an independent company that is only renting space in the studio. Such transience results in members being able to meet more fellow members, having only time to know one another casually but not like *family.*

PREPARING FOR PROMOTION

You are now a member of the Guild and an employee of a studio. By now you are referring to that studio as *my* studio or *my* company. The gate guard who two months ago might have appeared so threatening now gives you a cheery wave as you approach the time cards and punch yourself in.

If your studio has a studio club, you can join its activities, which may include bowling, baseball, or family events, and meet other employees from other departments. You may also be invited to employee screenings of new films. The aforementioned credit union can provide you with a savings account, an IRA, or a loan in addition to an opportunity for automatic savings.

Your present position as an apprentice may be as *permanent* a one as you will get for some time, provided you are the kind of apprentice the post-production supervisor wants to keep on her staff. This means being a conscientious and efficient worker. This does not mean trying to do more than you are capable of doing in an attempt to impress everyone. Do not promise or suggest something at the expense of someone else, and do not promise things you cannot deliver. Do not let your eagerness to please or impress get in the way of the simple truth of the situation. If you cannot perform an assignment immediately as requested, explain why you cannot and when you can. Then it is the other person's choice to wait for you or to get another apprentice to do it.

Concerning promotion, you must be listed as an apprentice on the *Industry Experience Roster* for three years before being eligible for promotion to any other classification except to that of motion picture editor for which a member must have served five years. Exceptions will be made if, for example, no one on the roster is available for a particular position or if you happen to have special qualifications for the job that those available in the higher classification do not have.

All apprentices look forward to the prospect of being promoted to assistant editor. An apprentice must, first of all, thoroughly learn the different tasks performed at the beginning level as have already been described. This is not easy, but it is particularly difficult for the apprentice assigned principally to one duty, such as numerical slate coding. This apprentice must use any spare time to practice other apprentice tasks such as splicing and winding film on a flange. In fact, *all* apprentices should practice these tasks until they are as competent at them as any assistant. Being adept at all of the apprentice duties prepares you only partially for the next promotion. Then and only then, should you be concerned about learning the principal duties of an assistant, which you must know before you can confidently assume the responsibilities of working with an editor.

The apprentice assigned to a cutting room on a television movie, a miniseries, or a feature or a *floating* apprentice helping a number of assistants will have an obvious advantage and is already learning the rudiments of the assistant's position on the job. But how do apprentices in shipping and receiving, in M&E, or in the film library get this additional experience?

First, during the period in which you are perfecting your apprentice duties you should be acquainting yourself with the editorial staff. Learn which assistants are assigned to which editors and on what projects, and unobtrusively let them get to know you. Try to determine which assistants, particularly those on episodic television series, have the best reputations regarding their work. (The nicest people may not necessarily be the best assistants.) Finally, you must decide which of these assistants are willing to let you spend time with them as they work and be patient enough to answer all your questions. The logistics of the arrangements you make are vitally important. The time spent with an assistant is usually when the editor is absent or, if she is present, with the editor's approval and without disturbing her editing!

When can you spend such time with an assistant? There are occasional periods when your department may not be very busy; if your department head and co-workers are agreeable, and if it is convenient for the assistant, you might be able then to spend a short time in the cutting room. But you cannot do this very often. Another possibility is that the assistant who has gotten to know you may ask specifically for you when she needs some temporary help.

However, your chief opportunity for this experience will be before you clock in and after clocking out. For example if your work schedule is 9:00 A.M. to 6:00 P.M. with one hour off for lunch, you would be able to be with the assistant from 7:00 to 9:00 A.M. when she would probably be *popping* daily tracks and filing *trims*. After work ends at 6:00 P.M. you might observe the assistant breaking down dailies, lining up a sequence for the editor, writing up optical count sheets, or breaking down opticals and cutting them into the picture.

If you do not have a short attention span and if you maintain your eagerness to learn and are appreciative of her interest in you, then the assistant's enthusiasm will continue, and as her confidence in you increases the assistant will begin letting you do the above chores under her supervision. Be understanding when it appears inconvenient for you to be in the cutting room. Do not be overzealous or aggressive, and you will not wear out your welcome.

Do not begrudge those extra hours you have to spend learning the duties of an assistant, even if you may not get a promotion as soon as you would like. Eventually you will be assigned to an editor as an assistant. Can you imagine what a difficult situation it would be for you without the training and what an abrogation to your career it would be if the editor had to replace you? Knowing you can handle the new position will make the transition from apprentice to assistant an enjoyable and exciting experience for you.

CHAPTER 6

THE ASSISTANT FILM EDITOR

Your efficiency and reliability as an apprentice in shipping and receiving/coding and your sacrifice of many precious nonworking hours to prepare yourself to become an assistant has not gone unnoticed. Besides the assistant who has been instructing you and his editor, other assistants and editors have become aware of your progress and aptitude and, when they have needed temporary help, have specifically requested your services. This has brought you to the attention of the post-production supervisor who, when the opportunity presents itself, will not hesitate to promote you to assistant film editor—or so our ideal scenario goes.

Your promotion occurs in July at the beginning of production for a new television season. You and your editor have been assigned to the first episode of a new one-hour series. In the ensuing two to three weeks, two more teams, consisting of one editor and one assistant, will be assigned to your series. The three teams will alternate the editing of the episodes.

What changes are there in your life as the result of your promotion from apprentice to assistant? Your promotion immediately moves you to a higher salary bracket. The basic agreement between the producers and the Guild sets a basic assistant's salary at about $150. more per week than your apprentice salary with an increase every six months until you are finally earning about $240. more than you were getting as an apprentice. You will be guaranteed pay for 45 hours weekly, which means that 5 hours of overtime is included in your weekly salary. You begin earning additional overtime *after* you have worked 45 hours.

There will also be an increase in your Guild obligations such as an additional $500. initiation fee and of course the approximately eight percent of your new salary for your quarterly dues. You presently have six assistants representing you on the board of directors of the Guild. Having been a member of the Guild for at least three years, you are now eligible for election as an officer as well as a director.

Why have you been hypothetically assigned as an assistant on an episodic series instead of a TV movie or some other project? And why has that area been selected as the major concentration of this text? There are several valid reasons. I have already mentioned that episodic television provides the greatest employment opportunities for editorial personnel and that consequently that is where most apprentices are first promoted to assistants. In addition, most (but not all) of these series present the greatest challenges to an assistant; the general concept is that, if an assistant can handle a difficult series, he can handle any project. Michael Kahn, one of Hollywood's foremost film editors,* said that the chief prerequisite that an assistant must have to be hired for one of his features is to have previous *television series* experience. ''I have confidence in assistants out of television,'' said Kahn, whose first editing credits were 130 episodes of ''Hogan's Heroes'' in a six-year period, ''because I know they will be fast and accurate.''

However, be aware of the factors that make the editing of a series more difficult than that of a television movie or feature film. Remember that a one-hour series has three teams, each with an editor and an assistant working on alternate episodes. You and your editor thus will be doing the first episode, the fourth, the seventh, and so on.

It may be a *difficult* series such as one with lots of action—chase sequences with cars, motorcycles, and planes—one with many optical effects. An action film is more difficult than a dialogue one because there is considerable location shooting with more takes being shot. This usually results in more troublesome dailies for the assistant. In addition, action sequences require more consideration and creativity on the part of the editor, which takes more time.

The shooting schedule for most episodes is about 7 working days; the shooting on the next episode begins immediately so that on the fifteenth working day following the last day of shooting on your first episode, your second episode begins. This, in turn, means that the editing of an episode must be completed within 21 working days from the first day of shooting so that both you and your editor can be ready for your next episode. Assume that each one-hour show has about 45 minutes of basic show time (*action footage*), the remaining 15 minutes being commercials, teasers, trailers, and other format material. Without consideration for music, sound effects, dubbing, cutting in opticals, and negative cutting you and your editor will have to finish three episodes, or 135 minutes of editing, within 63 working days (approximately three months).

Comparatively, the average feature, which runs less than 120 minutes, will take six to nine months to edit, two to three times as many months as you and your editor will work on a greater quantity of film. In all fairness, it should be pointed out, however, that on features much more time is spent by directors and producers in decision making, in experimenting with editing alternatives, and in refinement of the editing. Also, some features have release dates that limit their production time, bringing this

*Michael Kahn, Stephen Spielberg's principal editor, won an Oscar for *Raiders of the Lost Ark* and has won or been nominated for many other awards by the Motion Picture Academy, the American Cinema Editors, and the British Academy.

comparison somewhat closer. But the fact remains that all series have a *weekly* air date. In addition, unlike a feature, an episode of a series must total a set amount of footage that must be strictly adhered to. This places you and your editor under relentless pressure.

Other factors, such as script or shooting problems, network changes in air dates, or the shifting of episodes in the schedule, can increase this pressure. Under normal circumstances a difficult show might require consistently long working days of 12 hours (including a lunch period) with some days extending to 14 to 16 hours. Other factors might also extend the working week to six or seven days.

It is obvious how this will restrict your time for a private life, which will necessarily become *secondary* to your job. An apprentice assigned to a cutting room would be affected as well but not an apprentice in shipping and receiving. Now, as an assistant with new and very specific responsibilities shared with your editor, priorities have changed.

This sort of work load will be difficult for anyone with whom you have a close relationship, and you will find yourself forced into making many regrettable sacrifices. Do you have enough *passion* for this profession to make all the sacrifices, the hard work, and the sheer struggle for survival in this very competitive industry worthwhile?

Of course, not all series have such horrendous schedules. In fact, you might be fortunate and be assigned to an easy series (which still may require a 10 hour day). But even so when you are confronted with episode upon episode, you will have to be a first-rate assistant. In focusing the remainder of the text on the duties of the assistant editor on an episodic series, I believe I will be offering you the best general background for any assistant's position you may obtain in the future.

SETTING UP THE CUTTING ROOM

It is a sunny California day in early July. As has been previously outlined, you and your editor have been assigned to a new series. On your first day as an assistant the first episode of your series begins shooting. Your primary duty is to set up the cutting room because, even though your room may have been in use only a month or two ago, it is probably empty of all equipment and most furnishings either for use elsewhere or for security reasons. Your room is only 10 × 12 feet, much too small for everything you need, so you will have to arrange it carefully for maximum efficiency. It is your responsibility to do so.

You must first, before you order any film, paper material, or smaller articles, arrange for the furnishings and larger equipment such as cutting benches, reel racks, moviolas, chairs, and *trim bins*. Meet with your editor and get his suggestions and specific requests. Where would he like to have his bench? What about the telephone, should it be on his or your bench or between the benches? And ask him whether he has any preferences regarding the equipment. Would he prefer a moviola with pedals on the floor or off the floor? And does he want a three-way or a four-way synchronizer and with how many sound heads? Does he have other requests? You may not be able to get him everything he wants, but you should try.

Table 6–1. Setting up the cutting room

General room items
2 benches with chairs
Bench racks
Overhead lights and counter lights
Reel racks
Trim bins
1 telephone

Other items
Trim boxes
Reels
Extra 40-watt bulbs
2 butt splicers and spare blades
Picture and track splicing tape
Scratch pads and tablets
Pencils and pens
Legal- and letter-sized manila envelopes
Red and black brush pens
Grease pencils
2 tape dispensers and transparent tape
1- and ½-inch masking tape
4 three-ring notebook binders
2 clamps
Paper clips
Rubber bands
Gloves
Rolls of leader
2 wastebaskets
2 scissors*
2 magnifying glasses*
2 scribes*
2 network formats

Specific items for editor
Moviola with bag, light well, and tray
Extra moviola bulb
Bench rewinds (left one having swivel base and drag)
Three- or four-way sync machine
Banners missing:
  Insert
  Scene
  Hold frame
  Stock
Commercial banner for TV
Aperture matte*
Blowup chart*
Amplifier with 1 to 3 sound heads

Specific items for assistant editor
Bench rewinds
1-inch camera tape
Two-or three-way sync machine with amplifier and 1 sound head
Male flange
Female flange or adapter*
Tabs
Earphones*
Three-hole punch*
Webril® wipes
Forms:
  Code sheets
  Lab orders
  Sound orders
  Count sheets
  Screening records
  Footage chart
  Footage reports
  Stock reports
  Store requisitions
Lightboard or penlight*
Batteries
Clipboards

Optional
Carpeting
Easy chair and/or couch
Locker or coat rack
Roll rack
Directories*
Wall calendar*
High stool
A couple of single-edged razor blades*
1-inch ruler*
Paper fasteners
First-aid kit*
Screwdriver*
Pencil sharpener*
Film horse or roll devise*

*These items are generally not supplied by the film company. Individuals must supply them for themselves.

Your editor wanders off to the set, if the show is being shot at the studio, or to the producer's office to get a copy of the script to read. Having learned from the post-production supervisor's office how to order and obtain everything you need, you can now sketch out the room so that you will be able to direct the delivery men where to place the furnishings and equipment. These will include two cutting benches and bench chairs, bench rewinds, a moviola with large attachable bags, a telephone, a high stool (for a visiting director or producer), and, depending on space and availability, several reel racks, two or three large trim bins, easy chairs or a couch, a locker or coat rack, and carpeting.

You will have to pick up the required film, paper material, other supplies, and smaller equipment yourself unless you can get some help from an apprentice. Because of the many items required for a cutting room, an assistant should maintain a permanent list of them for future use. To get you started, a suggested checklist is provided here (Table 6–1) to which you may make deletions or additions as necessary. This list reflects minimum requirements, particularly for the assistant. Should you be able to get a moviola for yourself or a three-way synchronizer instead of a two-way, so much the better. While most of the items listed are self-explanatory (and others will be clarified as you learn the duties of an assistant), *banners* and *leaders* should be explained to you here.

BANNERS AND LEADERS

Banners are picture film specifying material not yet shot or ordered—insert missing; scene missing; stock missing; and *hold,* or *freeze, frame* (an optical effect that is described in Chapter 15). The editor will edit these in the picture wherever required until the actual film is received. Film for banners, as well as leaders, was ordered from the lab by the post-production supervisor and is kept with other supplies in thousand-foot rolls. You should wind off about 100 feet of each banner from these rolls and split this into 50-foot rolls, placing four rolls of each type on your editor's bench and keeping the other four on your bench as spares. (Commercial banners will be discussed in detail later, but briefly, they are used to indicate where commercials will ultimately appear in the finished product.)

Leader, *blank colored film,* may have been ordered in several forms, principally in yellow or grey. Like picture, it has an emulsion side and a cell side. You should start off with several thousand feet of the leader that is supplied. Break it down into 100- or 200-foot rolls, whichever your editor prefers for ease of handling, and place a couple of rolls on his bench. Leader is used chiefly at the head and tail of dailies and cut (edited) picture to identify and protect picture and sound. It is also used as temporary fill in the sound when there is no satisfactory track provided.

Whenever leader is used with picture, it should be like picture itself, *emulsion up, cell down,* meaning that the emulsion side is on the top, or facing, side while the cell is on the underside. When used with track, leader should be flopped over so that the cell side is up and the emulsion side down.

A *film horse* (Figure 6–1) is most often used by some sound and music editors rather than a picture editor. Should your editor possess one, however, position the horse on the extreme left side of his cutting bench beside the left rewind. Place one

Figure 6–1. Film horse.

reel of leader, emulsion up for picture, on one of the spindles and another reel of leader, cell up for track, on another spindle.

There is also salvaged black and white reversal film that may be used as leader only with track, never with picture, and not as the identifying head or tail leaders of a reel. When used as fill, reversals should be *tails up,* tail to head, as opposed to *heads up* meaning the beginning and end of the reversal picture are in reverse positions. Also, since B&W reversal is only used as leader for track, it should be flopped *cell up.*

DAILY LEADERS

When you have the room completely equipped and supplied, you should borrow your editor's script or obtain a copy from the producer's office. Using the information on the title page, make up half a dozen or more daily leaders in preparation for the dailies you will have to assemble the following day. It is better to have more than you need. Everything you can possibly do your first day should be done so that nothing will interfere with your concentration on dailies the next day.

Place two leaders in your synchronizer, leader for picture in the first gang *emulsion up* and track leader in the second gang *cell up.* There are five *primary* items you must inscribe on daily leaders (Figure 6–2):

1. *Production number:* Each episode has its own number such as 18201, 18202, and so forth. This number is normally the first item to be written on the leader, reading left to right.
2. *Production title:* Use the series title; episode titles are rarely referred to in postproduction and usually not even indicated on screen. But, if you wish to take the time, they can be included on leaders with the series title.
3. *Reel number:* Begin numbering, of course, with Reel 1.

Figure 6–2. Daily leaders.

4. *Code number:* Assuming that you are using the standard code system, I suggest that, since this is your first episode, you mark the code for your first reel A–1000, the second reel A–2000, and so forth. This coding will continue in numerical sequence from day to day. The tenth reel will be A–0000 and the eleventh reel, AB–1000, and so on with a constant *A* identifying the first episode. You would code your second episode with *B*, the third episode with *C,* and so forth. Using this kind of identification will avoid confusion when editing is being done on more than one episode. (At the very head end of the leader such extras as *Picture-Head* and *Track-Head* and date can be added, if you wish to take the time. However, these extras are usually reserved for the longer projects on which dailies are not broken down so quickly.)
5. *Start mark:* Make an *X* on a frame corresponding to both the picture and track at least six feet from the head end of the leader. This will provide enough film for the projectionist and the apprentice coder to thread their machines. Using the frame indicator and notches on your synchronizer (see Figure 4–5), make certain that your start mark is positioned in frame. Inscribing short lines or arrows on either side of the *X* will make the start mark even easier to locate.

These daily leaders should total about 10 feet in length: 6 feet from the head end to the start mark and then 4 feet to the end of the leader where the first take on the reel will begin. Be sure to cut the ends of the leaders exactly even to the sprocket *on the frame line.*

Remember to inscribe your leaders with *black* brush pen on picture and *red* brush pen on track. It is suggested that, instead of writing on the film itself, you write on masking tape or white tape attached to the head of the leader and at the start mark. Because continually reusing your daily leaders will save you considerable time and effort, using tape will prove to be more durable and efficient. Also, since the production number and title and start mark will remain constant, place an *extra* piece of tape only over the area where you indicate the reel and code numbers. When the dailies have been *broken down* or disassembled for editing, this extra tape can easily be removed and replaced with a fresh tape so that only the new reel and code numbers have to be rewritten.

Some studios require that a *focus chart* precede the first take of each daily reel so that the projectionist can adjust the focus without interfering with the picture. A roll of the film focus chart can be obtained from your post-production supervisor. You will have to add about 10 feet of this film to the tail of your picture head leader and of course insert equivalent leader to your sound to maintain sync.

One more important matter with regard to dailies: Ask your editor if it will be acceptable to assemble your dailies in *shooting sequence,* the order in which the takes were shot. Since this is the manner in which you will receive the picture and track, it is the most convenient procedure for you. Your editor may approve or he may advise you to do your dailies in *numerical scene sequence,* which sometimes requires considerable shifting of takes to get them in proper order and means that you must take much more time to do *(sync)* your dailies. You must know the correct procedure so that you will know how to proceed with your dailies the following day. This is discussed in detail in Chapter 8.

When you have completed everything you can about supplying your cutting room and preparing for dailies, with the approval of your editor you might visit the set if your company is shooting on the lot. Do not make a habit of hanging out on sets. Only visit a set on necessary business and with your editor's permission. On the set, introduce yourself to your most important contact on the crew, the *script supervisor.* Ask her if the *script notes* for each day might be included with the mixer's *sound reports.* As you will learn in the following chapters, the accuracy of the script supervisor's records and their coordination with those of other members of the crew will be of inestimable service to you and your editor.

CHAPTER 7

THE MAKING OF A FILM

Before you begin your first day of dailies, you should be familiar with all the steps taken from the beginning of shooting until the completed print from the perspective of the role played by you (the assistant), your editor, and the apprentice.

EVOLUTION OF FILM

In very brief, this is what happens:

1. 35mm film negative is loaded into the camera magazine. ¼-inch tape is placed in recorder.
2. Shooting begins.
3. Negative goes to lab for processing; only circled takes are printed. Tape goes to the sound department. Only circled takes are transferred to 35mm magnetic track.
4. 35mm track dailies are picked up by you or by an apprentice.
5. Tracks are *popped* by you (sound of the clapsticks is marked, enabling you later to synchronize track to picture).
6. Picture dailies are delivered to shipping and receiving and either picked up by you or delivered to the cutting room by an apprentice.
7. Dailies are synced by you.
8. Dailies are viewed by the editor, the director, and the producer.
9. Dailies are coded by an apprentice.
10. Dailies are broken down by you.

11. Editor begins cutting sequences.
12. Shooting, dailies, and editing continue.
13. Shooting ends. Dailies end the following day.
14. Editing continues. (Cut sequences are assembled.)
15. First cut is screened for the director. Editing changes, screenings, and final cut on most episodic series only for the producer.
16. Spotting sessions for music and sound effects are done.
17. Opticals are ordered by you.
18. Picture "turned over" to M&E. Music is composed, scored, and edited. Sound—sound effects, dialogue, foley, and ADR—are done.
19. Opticals are cut in by you.
20. Negative is cut by negative cutter.
21. Dubbing takes place.
22. Answer print (composite print with optical track) are made.
23. Film transfers to videotape are made.

SHOOTING PROCEDURE

On your first day as an assistant you have been busy setting up your cutting room, getting all the equipment and supplies you will need, and preparing for dailies that will begin the following day. Meanwhile back at the set we will assume your company is using the more common standard scene/take system rather than the numerical slate system. The script supervisor assigns scene numbers to each scene and to any angles (or *coverage*) of that scene. These numbers will not necessarily correspond to the scenes as numbered by the writer in the script.

Take, for example, a scene in a classroom with students listening and responding to their instructor. The scenarist has begun this as Scene 35. At various places where the writer anticipates a radically different camera setup, such as a student entering or leaving the classroom or a student joining the instructor at the blackboard, she may indicate a new scene number, finally concluding the scene with Scene 40.

The script supervisor will also begin with Scene 35 but then she will identify every change of angle that is shot as follows:

Scene 35	Master or full shot
35A	Pick up (P.U.) master
35B	3/shot (3/s) Tony, Phyllis, and Claire
35C	2/s Bob and Helen
35D	Close up (C.U.) Helen
35E	C.U. Bob
35F	Reverse master, over class to instructor
35G	Medium (Med.) reverse shot favoring (fav.)
Scene 35G (cont.)	Tony, pan with him to door exiting
35H	Med. shot (Med/S) Steve enters, pan to seat
35I	Side angle fav. Bob, pan to blackboard
35J	Med. C.U. instructor
35K	P.U., Med. C.U. instructor to 2/s with Bob

Thus, Scenes 35 through 35K replace script scenes 35 to 40. In addition, each *take* of each scene will be numbered numerically until the director is satisfied with a take and orders it to be printed. So, the *printed* takes of the above might be: 35–Take 3, 35A–5, 35B–1, 35C–3, 35D–2, and so forth.

Shooting begins with the first take of the first angle, usually the master shot. An assistant cameraman will hold the *clapsticks,* or *slate,* with the hinged board open, directly in front of the camera. The slate identifies this shot as Scene 35, Take 1, and also lists the production, director, director of photography, shooting date, and other technical information.

The camera starts shooting the slate. The *mixer* (not to be confused with a *mixer* on a dubbing stage), also called the *soundman* or *recordist,* begins recording on ¼-inch audio tape and announces the scene/take number that has been given to the mixer by the script supervisor. When the speed of the tape in the recorder matches camera speed (90 feet per minute), the mixer says "Speed." The camera operator, having checked for correct focus and otherwise ready, calls "Marker" or "Mark it," prompting the assistant cameraman to bang the clapsticks together. Then the director calls, "Action!", and the master scene commences. Unless there is some unexpected interruption this preliminary procedure takes only a few seconds. Time is precious on a set.

At some point the director may stop the scene by calling "Cut!" because of performance or technical problems. The crew then starts over again with a new take, in this case Take 2. Even if it was a complete take and a fairly good one, the director may decide it can be improved and the scene would thus be reshot again as Take 3. Even though the director orders Take 3 printed, she is not completely satisfied with the last part of the scene. In order to save time and energy instead of redoing the entire master shot, only the end section of the take is reshot. It is numbered Scene 35A and as before will have numbered takes, beginning with Take 1 until a printed take is ordered. Since 35 was the master and 35A is a piece of the same angle, it is identified by the script supervisor as a P.U. of the master.

The same general procedure will be followed with each take and new angle. The bang of the clapsticks on film and its corresponding recorded sound provide you with a definite visual and audible point at which you will be able to sync up picture and track the following day.

Concurrently with the shooting, certain records are kept by crew personnel and later by your sound department and lab; these records are vital to an efficient cutting room as you will see.

DAILY FORMS

The script supervisor, an assistant cameraman, and mixer keep their own records of all the takes shot. It is the responsibility of the script supervisor to make certain their records correspond particularly in regard to those takes ordered printed. The printed takes are identified by being circled, and therefore are commonly referred to as *circled* takes. A scene would be listed as illustrated in Figure 7-1. Take 2

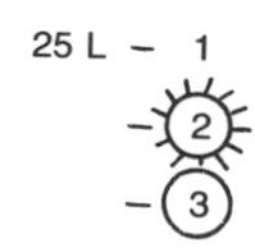

Figure 7–1. Circled takes.

was originally circled; then the director decided to shoot another take, 3, liked that better, and take 2 was cancelled by *marking* the circle and take 3 was circled and printed.

These important reports and the people responsible for them are described below.

Script Supervisor

The script supervisor has many responsibilities. She monitors the dialogue; watches for any possible mismatch in the action, wardrobe, or props between or within scenes or takes; and advises the director of any incongruity. She times each take, records the camera angle, and describes the action or subject. She also assists performers' rehearsing by acting as prompter. She is often required to read the off-screen lines during shooting.

An efficient script supervisor can be invaluable to the editors. She sends the editor copies of the *lined script* pages that have been shot each day. On each page of the script that has been shot she should record all the printed takes with a brief identification, and by *lining* each take she should indicate whether the inclusive dialogue or piece of action is on screen or off screen. Such accurate and complete information on these pages will greatly assist the editor. Also included in the lined script on the reverse side of the script pages is a *log* of *all* the takes shot. Those takes *not* circled are referred to as *outtakes,* or *B negatives.* The complete original lined script is sent to the editor at the completion of shooting (see Chapter 10, especially Figure 10–3, for a more detailed description).

The script supervisor also makes a daily list of *only* the circled takes with brief descriptions of each take. Therefore, these lists are sometimes referred to as *script notes* but are more commonly called *one-liners.* Some companies may reverse the terminology by referring to the one-liners as *logs* and the lists in the script of all the takes shot as script notes. Confusing, isn't it? But you were warned that as your work takes you from one company to another you must be prepared to adapt to different procedures as well as diverse terminology. For the purpose of this text one-liners are *script notes.* They are sent to you, the assistant, as a guide for your dailies because only circled takes comprise the dailies (see Chapter 8, especially Figure 8–1).

As suggested in the previous chapter, you should make it a top priority to introduce yourself to the script supervisor. She will be your most important contact on the crew. When you introduce yourself, you can also ask her if she might enclose her script notes with the sound report and sound that is sent to the sound department so that you can receive it as early as possible each day.

Probably the script supervisor's most valuable service to you, though, will be the coordination of her records with those of the mixer and the assistant cameraman.

Mixer

The mixer sends the ¼-inch audio tape with a *sound report* to the sound department. She lists *all* the takes shot, and this list should coincide with those on the script supervisor's log.

Assistant Cameraman

The assistant cameraman's *camera report,* which must also correlate with the other reports, is sent to the lab each day along with negative to be processed. It also contains *all* the takes shot. A copy of the camera report is sent to the assistant editor.

Sound Department and Lab

The sound transfer person in the sound department transfers all *circled* takes indicated on the sound report from ¼-inch tapes to 35mm mag tracks. These are your sound dailies, and they are listed on the *sound transfer report.*

The lab prints *only* the circled takes designated on the camera report. These picture dailies are listed on a *lab report* that is sent to you along with the daily prints. The uncircled takes, our outtakes or B negatives, are not printed.

In summary, the script notes, sound transfer report, and lab report list *only* the circled takes. The other two forms you receive, the sound report and camera report, list *all* the takes. The lined script that is delivered to your editor also logs all the takes. We will examine each of these forms in more detail as you, the assistant editor, work with them.

EXCEPTIONS

It has already been emphasized that procedures may vary with different companies. As we proceed I will note some of these exceptions as well as those likely to be found on film projects dissimilar to the average episodic TV series.

When you are using the numerical slate system instead of the scene/take system, the first angle shot on the first day of shooting begins with Slate 1, Take 1. As with the scene/take system since *takes* are numerical for each slate if shooting continued on the *same* angle, the next takes are slated 1–2, 1–3, and so forth. The next angle shot begins with slate 2, take 1 and so it continues numerically through to the final shot on the final day. Refer to our earlier hypothetical production where the first shot of the first day is Scene 35. Instead of the scene/take system if we had used the numerical slate system, the circled takes would be slate 1, take 3, 2–5, 3–1, 4–3, 5–2, and so forth. The script supervisor, in this case, will make cross-references on her records so that you and your editor would know that Slates 1 through 12 comprise all of the coverage for Scene 35 or script scenes 35 through 40.

Another exception that should be further explained is the role of the director and producer in regard to post-production. The length of time directors remain on their projects varies with each project according to their agreements with the producers or companies. There are relatively few directors who retain full control of the editing and remain through dubbing and the answer print. Producers assume final control on the majority of projects.

This is particularly true in episodic television where directors are assigned to a single episode, screen a first cut, and perhaps view the editing changes before going on to direct episodes of other series. Producers then complete the supervising of the editing of practically all one-hour programs.

CHAPTER 8

DAILIES: ORGANIZATION AND PREPARATION

It is the second day of shooting—your second day as an assistant editor and your first day of dailies. *Dailies* are those takes shot the previous day that the director ordered printed.

Scripts are not shot in script sequence—from beginning to end. Instead, scenes that have the same or contiguous locations are scheduled together, regardless of where they occur in the script, so that time and money can be saved. There are three ways in which you might assemble your dailies: in shooting sequence, in similar scene sequence, or in numerical scene sequence.

Shooting sequence is the easiest way for you to *build* your dailies. The circled takes are printed and transferred in the order in which they were shot. You simply assemble them in that order.

Similar scene sequence. A good assistant should automatically combine similar scenes with shooting sequence. To illustrate, let us suppose the first circled takes shot were in shooting order: 39–2, 39A–1, 39B–1, 24–1, 25–4, 25A–2, 25B–3, 25C–2, 46–2, 30–1, 25D–3. You would move 25D–3 to after 25C–2 and before 46–2 so that all angles of the same scene were together.

Numerical scene sequence. Some directors, producers and editors require that the assistant screen the dailies in numerical scene sequence. This is obviously more difficult and time consuming than the other assembly orders. You have to rearrange the takes listed above as follows: 24–1, 25–4, 25A–2, 25B–3, 25C–2, 25D–3, 30–1, 39–2, 39A–1, 39B–1, 46–2.

Do not confuse numerical scene sequence with the numerical slate system that was discussed in Chapter 7. The former refers to assembling dailies, the latter to slating the takes. Numerical scene sequence will be discussed further later in this chapter.

Since the circled takes on the sound reports are transferred overnight from the ¼-inch tape to 35mm track, you come to work early, between 7:00 and 8:00 A.M., so that you can pop the tracks (mark the beginning sounds of the clapsticks) before the picture dailies arrive from the lab at about 9:00. Pick up the tracks, the sound transfer report, a copy of the sound report, and if possible the script notes from the sound transfer department.

When you get to the cutting room, anxious as you may be to begin popping the tracks, first compare the takes listed on the sound transfer report with the takes indicated on the script notes. Your script notes are preeminent so if it lists a take, except for a *MOS* (silent) take, that you do not have on your sound transfer report, then that take must be ordered. You can order any missing tracks by phoning the sound transfer room. By the time you have finished popping the other tracks the missing ones should be ready for you to pick up in plenty of time to prepare them for the picture.

SCRIPT NOTES

These notes (Figure 8–1) from your script supervisor are your *bible* in regard to what takes should constitute your dailies. The takes are listed in shooting order, and in the sample script notes we can see immediately that the scene/take system is being used rather than the numerical slate system.

In addition to the takes the script notes indicate the running time of each take as timed by the script supervisor and a brief description of the scenes. Note that three of the takes are *MOS,* (mike off screen *or* mid-out sound). In other words, they were shot silent, so you do not receive any tracks for those scenes.

Other useful information is usually included when it is appropriate, multiple cameras, no slate, second announcement, end slate, and so forth. The sample script notes list a *WT 25* that is a *wild track* for scene 25. This means sound was recorded without picture so you do not receive a matching picture.

Finally, as I mentioned in Chapter 7, I have tried to base my procedures and terminology on common usage but there appears to be uncommon disagreement from company to company about the designation of what I identify as *script notes.* Some companies refer to them as *logs* or *daily reports.* The most common term seems to be *one-liners.*

SOUND REPORT

All takes, outtakes and circled takes, are listed on this report (Figure 8–2). In the sample sound reports the mixer has indicated such information as *E.M.* (end marker) and *no ann* (no announcement). An end marker (or end slate) is placed at the tail or

Columbia Pictures Television

A division of Columbia Pictures Industries, Inc.
A subsidiary of The Coca-Cola Company

CUTTER'S SCRIPT NOTES

SHOW TITLE: "Murder, Anyone?" PROD. No. 8191

SCRIPT SUPERVISOR: M. Sharon SHOOTING DATE: 10/4

SCENE No.	SHOT No.	TIME	DESCRIPTION
39	2	:17	Int. Hallway-Kirkwood/Ryan to apt. door
39A	1	:10	o.s. Golman at door
39B	1	:13	o.s. Kirkwood/Ryan
24	1	:34	Int. Hallway-Kirkwood/Ryan/Bellamy to apt
25	4	:44	Int. apt - men enter
25A	2	:46	Master) p.u. Bellamy exits, Marshall enters)
WT 25	1		Marshall's off-screen dial.
25B	3	:38	c.u. Kirkwood
25C	2	:35	c.u. Ryan
25D	3	:45	c.u. Bellamy
25E	3	:20	p.u., c.u. Bellamy - exits apt.
25F	4	2:50	med/s Kirkwood/Ryan at door to 2/s
25G	1	2:55	med. to c.u. Kirkwood
25H	2	2:52	med. to c.u. Ryan
25I	1	:15	p.u., c.u. Kirkwood
25J	4	2:46	Master - diff. angle
25K	1	:05	c.u. Marshall
25L	3	2:22	MED/S Marshall
25M	1	:10	MOS Pan trophies R-L, L-R, mantel
↓	2	:08	MOS med. c.s. trophies on mantel
25N	1	:05	MOS c.s. trophy onto bedstand (3 takes)
46	2	:25	Int., Daniels/Ryan in office
30	1	:22	Int., Daniels on phone in office

ML-1965 7/85 R

Figure 8–1. Script notes.

PRODUCTION SOUND REPORT

Roll No. 14
Page 1 of 2

TWENTIETH CENTURY FOX FILM CORPORATION
10201 W. Pico Blvd., West Los Angeles, CA 90035
Phone 213-277-2211, Re-Re Ex. 2873, Maint. Ex. 2431

Production No.: 8191 Date: 10/4
Title: "Murder, Anyone?"
Mixer: Carlson Cable: Randall
Boom: Stevens Location:

PRINT CIRCLED TAKES ONLY

Scene No.	Take	Notes	Scene No.	Take
39	1			
	(2)	40		
39A	(1)	30		
39B	(1)	30		
24	(1)	70		
25	1			
	2			
	3			
	(4)	85		
25A	1			
	(2)	90		
WT 25	(1)	75 Marshall off-screen dial		
25B	1			
	2			
	(3)	80		
25C	1			
	(2)	75		
25D	1			
	2			
	(3)	85		
25E	1			
	2			
	(3)	45		
25F	1			
	2			
	3			
	(4)	290		
25G	(1)	295		

IF LAST ROLL OF DAY – CHECK HERE ☐

A

PRODUCTION SOUND REPORT

Roll No. 15
Page 2 of 2

TWENTIETH CENTURY FOX FILM CORPORATION
10201 W. Pico Blvd., West Los Angeles, CA 90035
Phone 213-277-2211, Re-Re Ex. 2873, Maint. Ex. 2431

Production No.: 8191 Date: 10/4
Title: "Murder, Anyone?"
Mixer: Carlson Cable: Randall
Boom: Stevens Location:

PRINT CIRCLED TAKES ONLY

Scene No.	Take	Notes	Scene No.	Take
25H	1			
	(2)	295 E.M.		
25I	(1)	35		
25J	1			
	2			
	3			
	(4)	280		
25K	(1)	20		
25L	1			
	(2)	248		
	(3)	240		
46	1			
	(2)	50 NO ANN		
30	(1)	45		

IF LAST ROLL OF DAY – CHECK HERE ☒

Form P.S. 725 Rev. 1/78

B

Figure 8–2. Sound reports.

end of a take instead of at the head or beginning where the shot is usually identified. *No announcement* is when the sound technician has not announced the slate number. Multiple cameras, second announcements, or any other information that may be important should also be noted on the sound report.

The mixer gives a footage on the printed takes and assigns succeeding roll numbers to each ¼-inch tape used, recording them on the upper right-hand side of the reports. In Figure 8–2 the roll numbers are 14 and 15 and on the following day the mixer would begin with 16. Had it been the first day of shooting of your episode, the sound technician would have begun with rolls 1 and 2 and continued on the second day with roll 3.

After it has been recorded, each roll is placed in its cardboard container; identified on the side with production title, production number, roll number and date shot; and sent to the sound department.

SOUND TRANSFER REPORT

The sound transfer person, using the sound report as a guide, transfers *only the circled takes* from the ¼-inch rolls to 35mm magnetic track and lists them on the sound transfer report (Figure 8–3). The tape roll numbers (14 and 15) are also noted. The

SOUND TRANSFER RECORD

COMPANY "Murder, Anyone?" DATE 10/4
PROD. NO. #8191 ORDER NO.
RECORDER NO. 13A START TIME
REPRODUCER NO. 2 FINISH TIME
FILM TYPE SS new base CHAN. TIME HRS. MIN.
ISSUED TO Editor ROLL NO. 14
OPERATION Dailies FEET
SPEC INST. George FEET ISSUED 1300′

TAKE NO.	COUNTER READING	TAKE FOOTAGE	REMARKS
25G-1	715	295	
25F-4	420	290	
25E-3	130	45	
25D-3	85	85	
25C-2	575	75	
25B-3	500	80	
WT25-1	420	75	
25A-2	345	90	
25-4	255	85	
24-1	170	70	
39B-1	100	30	
39A-1	70	30	RECEIVED BY
39-2	40	40	

PF 56

№ 36950

A

SOUND TRANSFER RECORD

COMPANY "Murder, Anyone?" DATE 10/4
PROD. NO. #8191 ORDER NO.
RECORDER NO. 13A START TIME
REPRODUCER NO. 2 FINISH TIME
FILM TYPE SS new base CHAN. TIME HRS. MIN.
ISSUED TO Editor ROLL NO. 15
OPERATION Dailies FEET
SPEC INST. George FEET ISSUED 1000′

TAKE NO.	COUNTER READING	TAKE FOOTAGE	REMARKS
30-1	965	45	
46-2	920	50	no ann.
25L-3	870	240	
25K-1	630	20	
25J-4	610	280	
25I-1	330	35	RECEIVED BY
25H-2	295	295	end marker

PF 56

№ 39323

B

Figure 8–3. Sound transfer reports.

report repeats such information as *no ann* and *end marker*. In fact, the transfer technician will add such information he might notice that the mixer neglected to record.

POPPING TRACKS

What is *popping tracks?* As explained previously, it is marking beginning sound of the clapsticks as a means of synchronizing sound to picture.

Having checked your transfer report with your script notes or, if for some reason you have not yet received the script notes, with your sound report, and if you are confident all the tracks are accounted for, you can then proceed with popping (or marking) the tracks.

Roll 14 tracks are split onto two rolls of 35mm single-stripe track on cores, one roll containing scenes 39–2 to 25C–2 and the other roll having scenes 25D–3 to 25G–1. Roll 15 has scenes 25H–2 to 30–1.

Assuming that your editor has agreed to let you assemble dailies in shooting sequence, you will not have to reshuffle the takes. Since the sound tracks usually come out of the transfer room tails out, you, as a beginning assistant, should rewind all three rolls onto reels to their heads. Take 39–2 will now be at the head of the first reel; 25D–3, at the head of the second reel; and 25H–2, at the head of Reel 3. The tracks should be emulsion (dull) side up.

Take the reel beginning with 39–2 and place it on the left rewind. Place an empty reel on the right rewind and lay the track onto the second gang of your synchronizer (see Figure 4–5), which is always reserved for sound. Built into the second gang is a sound head that will pick up the sound recording from the track and, via an amplifier, enable you to listen to it. Make certain that the sound head is plugged into the amplifier and that the amplifier is turned on. As you thread the track through the synchronizer, position the sound head on the sound stripe so that the head is clearly in contact with the stripe. Roll the track to your right, winding it onto the empty reel. There should be several feet of track with only a *sound tone* preceding the first take on each transfer roll. A sound tone is used by the sound transfer person to set a level on the sound before it is transferred.

As was explained earlier, the primary objective of popping tracks is to locate and mark the precise beginning sound of the clapsticks. The following suggestions will help you learn to pop tracks correctly:

1. In the learning stages pop your tracks from head to tail. After a few days you should be able to pop them from tail to head.
2. Except for the first take on each transfer roll, listen for the *drop-out* in sound, or blank track, that occurs between takes alerting you to the subsequent take. (In some cases the mixer may have recorded two quick tones or beeps at the end of a take and one at the beginning of the next take. These *signals* will help the sound transfer person when he is transferring the circled takes but are rarely included in the transfer and therefore cannot be used by the assistant. Nor are they needed to pop the tracks.)

3. Listen for the slate announcements, making certain that they correspond with those listed on the sound transfer report.
4. Listen for the mixer's pronouncement of *speed,* meaning that the sound has caught up with the speed of the camera (24 frames per second).
5. Next listen for the command *mark it* or *marker* that directs the cameraman to slap his clapsticks together.
6. Immediately after this order you should hear the sound of the clapsticks.
7. Finally, listen for the director's *Action!* to be sure there are no other clapsticks.

The speed with which you wind the track is important. If you wind too fast or too slow you will not be able to distinguish the above sounds so try to achieve an even pace. Do not slow down or stop until you have passed the clapstick or marker and proceeded to *Action!* thereby confirming that there are no further clapsticks. Without necessarily presetting the synchronizer footage counter at any particular footage, note the footage as you pass the sound of the clapstick so that you have a temporary reference. After you hear *Action!* you will need to return to a point a foot ahead of where you heard the marker.

Next you need to pinpoint the very beginning of the sound of the clapstick. When you are working with just a few feet of track, if you use the hand roller on your sync machine instead of the rewinds to move the track, you will find that you are better able to control the track in locating that ninety-sixth of a second that marks the start of the marker. You can also achieve the same results by holding the far edges of the track (near the balancing stripe) on either side of the synchronizer to move the track. But be sure not to touch the sound stripe. When you have located on the track the very beginning of the clapstick, mark in white grease pencil that point between the two sprockets closest to the middle of the sound head.

Having popped the clapstick to establish its beginning, move the track to your right onto the bench where you can write more easily; extend the mark (see Figure 8–4), and identify the take. Write distinctly, avoiding the sound stripe, and in large print so that you will be able to find the spot quickly when you are syncing track to picture. In our hypothetical scenario you have now completed your first take, scene 39–2.

You will use the same procedure in popping all your takes, although a few variations may be imposed on you. In your first day's dailies you have two variations, WT 25 and 25H–2 (with an *end marker*).

WT-25 is a *wild track* for scene 25. Referring to your script notes, you see it is "Marshall's off-screen dialogue." Wild tracks, or wild lines, may be either dialogue or background sound recorded without picture. An astute director who takes the time to do this may be eliminating the need to bring an actor back for post-production recording or may be saving sound effects personnel considerable effort in duplicating the background.

Since WT–25 has no picture, there should be only an announcement. Splice out the wild track, making your cuts in the *drop-out* areas before and after the take. Roll it up heads out, identify it, and, securing it with a paper clip or rubber band, place it temporarily on one of your bench shelves.

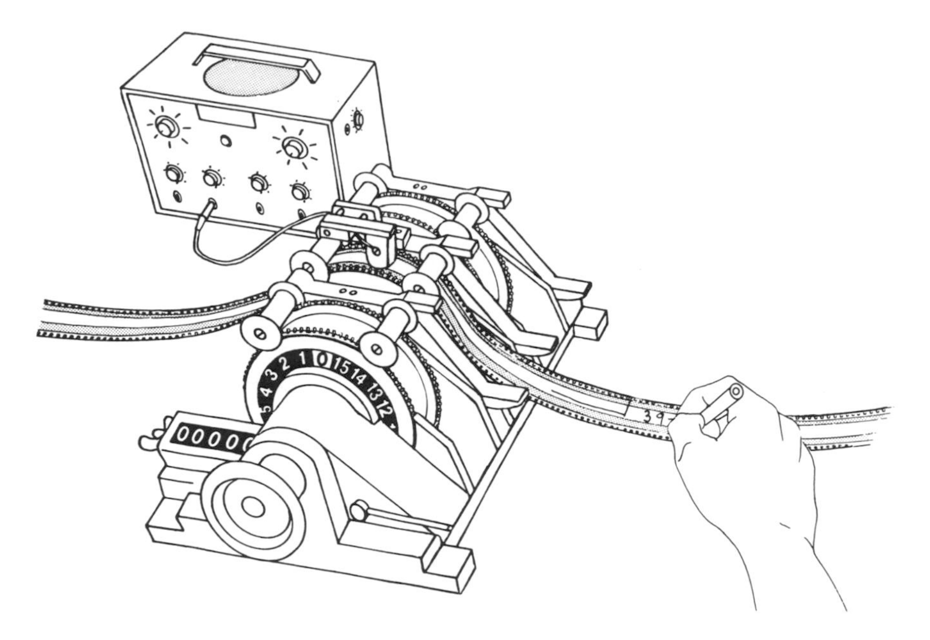

Figure 8–4. Marking the track.

The clapstick for 25H–2 is at the end of the take. For your own information later when you are syncing track to picture, note at the head of the take that it has an end marker (EM) or end slate (ES) at the tail. End slates are done either because the crew neglected to slate at the beginning of the take or for a particular reason having to do with the performers. For example, the harsh sound of the clapsticks will often frighten children or animals about to perform or it may disturb the mood or concentration of actors in a difficult scene so the clapsticks will be intentionally delayed until the end of the take.

Sometimes there may be more than one clapstick. The last one is usually the correct one to use but indicate the previous ones with question marks. If all your pops match up with all the picture clapsticks, you have achieved sync. If, for example, there is only one clapstick visible on screen, how would you know which pop the clapstick matches? You will probably have to use a moviola to be assured you have the correct sync. This is further discussed in Chapter 9.

When more than one camera is used to shoot a scene, it is called using *multiple cameras*. Each camera shoots from a different angle and has its separate clapsticks and film, but there is only a single sound tape for all the cameras. When the ¼-inch tape is transferred to 35mm, you will be able to hear the clapsticks of all the cameras on one track. However, the transfer room should give you multiple transfers, one after the other, so that you will have a complete track for each camera in consecutive order. Each clapstick will be preceded by an announcement identifying the corresponding camera, *A* camera, *B* camera, *C* camera, and so on. You will have to pop and identify each clapstick. If, for example, scene 75 is shot with two cameras, you identify the markers as 75–3 *A* on one track and 75–3 *B* on the other track.

Sometimes the clapsticks are difficult to hear or to distinguish. Just as they hit, someone on the set may sneeze or cough or a crew person might drop a board, making it difficult to determine which sound is the clapstick. This forces you to match track to picture in the moviola. This is called *lip syncing* and will be described in the next chapter.

When you finish popping all the tracks, rewind them to their heads. You can do as many as four at a time on the bench rewinds using a clamp to hold together the take-up reels. Your tracks will now be in shooting sequence and ready to be aligned with the picture.

It should take you only a few days to become so confident with popping tracks that you will make a couple of changes in the suggested procedure to increase your speed. First, having achieved an even pace in your winding, you will begin to zip through the action of each take until you hear the drop-out. Remember that your objective is not to listen to the dialogue or action but simply to mark the clapstick and identify it.

Second, instead of rewinding the tracks and popping them from head to tail as outlined above, you will begin to pop the tracks from tail to head by stopping at the dropouts between the takes and backing up for the clapsticks. This will save you the time of rewinding the tracks twice.

When the tracks have been popped, get any missing tracks you may have had to order from the sound department and pop them, being sure to identify them carefully. You need not put them into the track reels. Just put them in rolls on one of your bench shelves, and you will be able to integrate them into the proper place as you are syncing the picture.

When you go to the sound department to pick up the track order you previously phoned in, you should give them a confirmation order. This is an ordinary sound transfer order with *confirmation* written on it, indicating you are verifying the order previously called in and are not placing a new order. (Sound orders are discussed in Chapter 11.)

Each day you receive dailies you will follow this procedure for popping tracks. Rarely are you given any prior information on how many dailies or how many tracks you will have to pop the following morning. Since assistants want their tracks to be prepared before the picture arrives, most of them will pop their tracks between 7:00 and 8:00 A.M.

When this task is completed, there will be time to roll up trims, splice, or do other necessary chores. With all that done there is always a commissary where you can join co-workers for breakfast and conversation. This should certainly put you in a very positive mood, ready for the picture dailies that will arrive shortly.

NUMERICAL SCENE SEQUENCE

Should you have to assemble your dailies in numerical scene sequence, *break down* your tracks as you pop them. Only a little forethought is required. Rather than separating each individual take, you have to separate groups of takes. Refer back to your script notes or sound transfer reports and ask yourself how would you best prepare these tracks for numerical assembly.

As has been suggested, you will eventually be popping from tail to head and you will be able at the same time to break down the groups of tracks or individual tracks. This results in rolls being placed temporarily on your bench rack or reels on a convenient reel rack depending on the amount of film involved. Usually more than 400 feet of track or picture is best handled on a reel. Line up the rolls and reels in their respective places for the intended assembly sequence with the head pops visible for identification. The manner in which you break down and organize your tracks will mainly, with the exception of any silent takes, dictate how you will break down your picture dailies.

Beginning at the end of the first roll of roll 14 (Figure 8–4), 24–1 through 25C–2—splicing out WT 25—are retained as a group on a reel, leaving 39–2 through 39B–1 as a separate roll. The second part of roll 14, 25D–3 through 25G–1, remains intact on a reel. Roll 15 (Figure 8–5) results in 25H–2 through 25L–3 comprising a reel while 46–2 and 30–1 have to be individually rolled up and placed on your bench shelf. Be sure you know the beginning take on each roll or reel. They are now easily accessible for numerical sequencing beginning with 24–1 through 25L–3, 30–1, the 39s, and finally 46–2.

All major studios have sound departments convenient to editing rooms, but should you work for an independent company your tracks may be transferred by an independent sound house located elsewhere and you will have to depend on transportation to deliver your daily tracks. A missing track may not be obtainable in time for your screening, and you may be forced to run the corresponding picture silent, syncing it up later.

CAMERA AND LAB REPORTS

On a typical day, after you have popped the tracks and completed any other chores, pick up your copies of the *camera reports* (Figures 8–5 and 8–6) either from the camera department or from the post-production supervisor's office. Then you wait in shipping and receiving until your picture dailies arrive. Depending on the lab servicing your company, dailies generally arrive between 9:00 and 9:30 A.M. Always let your editor know where you are going to be, either verbally or by note, until he is familiar with your routine.

Picture is delivered from the lab tails out on cores, each roll in its own cardboard box. These cartons, which are approximately 10½ × 10½ × 1½ inches, will eventually be used as *trim boxes* in which unused film is filed. A *lab delivery slip* (Figures 8–7A, 8–8A, and 8–9A) is attached to each container to identify the production, roll number, and footage of the film enclosed. Note the date of delivery on the slip. The lab includes the date with the roll number so that D1005–147 in Figures 8–7A and 8–7B means October 5 and roll 147. Therefore, like the script notes and sound forms described earlier, these delivery slips as well as the camera reports and subsequent forms illustrated will be considered hypothetically *your* forms even though the dates do not coincide with your first episode in July. Also note that Figures 8–5 and 8–6 illustrate two different camera report forms—one from a studio and another from a lab. Realistically, you would, of course, receive all your camera reports from one company.

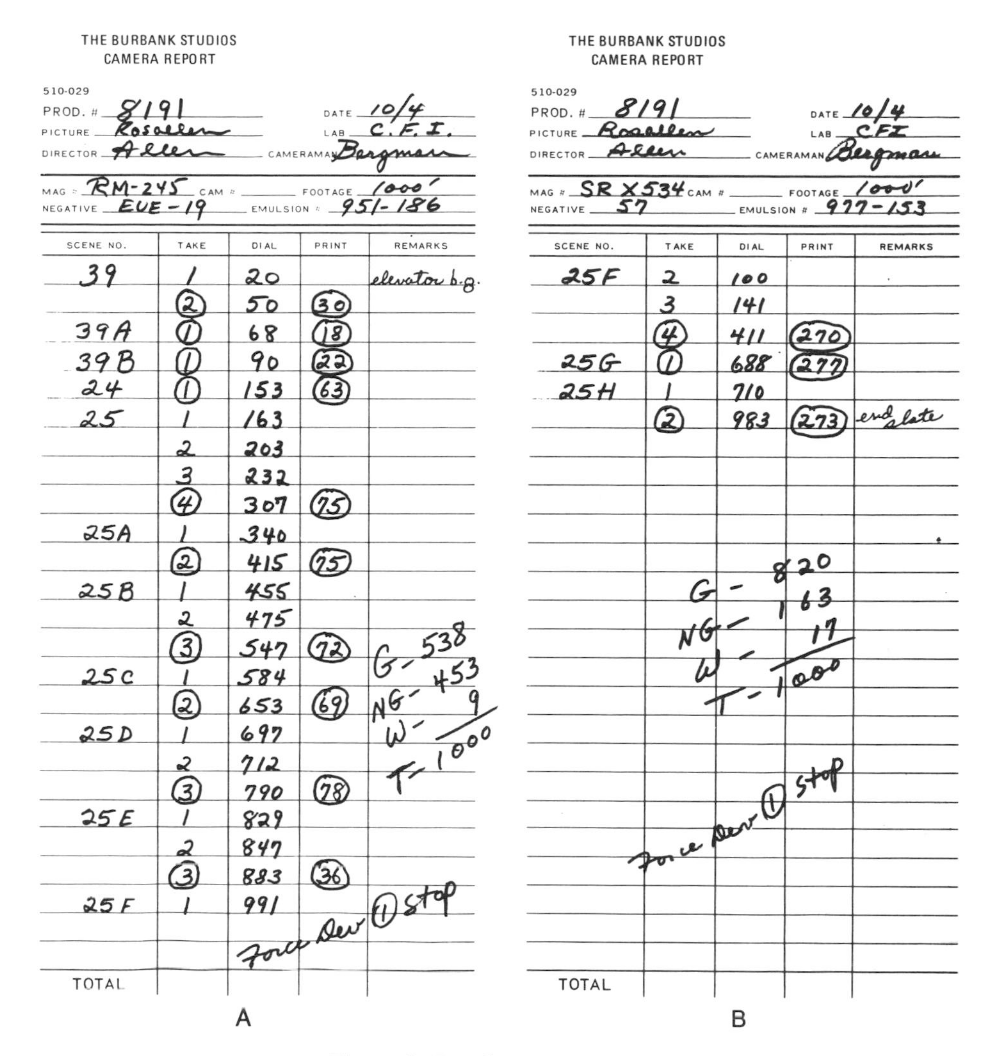

THE BURBANK STUDIOS
CAMERA REPORT

510-029
PROD. # 8191 DATE 10/4
PICTURE Rosallen LAB C.F.I.
DIRECTOR Allen CAMERAMAN Bergman
MAG # RM-245 CAM # FOOTAGE 1000′
NEGATIVE EUE-19 EMULSION # 951-186

SCENE NO.	TAKE	DIAL	PRINT	REMARKS
39	1	20		elevator b.g.
	(2)	50	(30)	
39A	(1)	68	(18)	
39B	(1)	90	(22)	
24	(1)	153	(63)	
25	1	163		
	2	203		
	3	232		
	(4)	307	(75)	
25A	1	340		
	(2)	415	(75)	
25B	1	455		
	2	475		
	(3)	547	(72)	G-538
25C	1	584		NG-453
	(2)	653	(69)	W- 9
25D	1	697		T-1000
	2	712		
	(3)	790	(78)	
25E	1	829		
	2	847		
	(3)	883	(36)	
25F	1	991		
		Force Dev (1) stop		
TOTAL				

A

THE BURBANK STUDIOS
CAMERA REPORT

510-029
PROD. # 8191 DATE 10/4
PICTURE Rosallen LAB CFI
DIRECTOR Allen CAMERAMAN Bergman
MAG # SR X534 CAM # FOOTAGE 1000′
NEGATIVE 57 EMULSION # 977-153

SCENE NO.	TAKE	DIAL	PRINT	REMARKS
25F	2	100		
	3	141		
	(4)	411	(270)	
25G	(1)	688	(277)	
25H	1	710		
	(2)	983	(273)	end slate
	G - 820			
	NG - 163			
	W - 17			
	T - 1000			
	Force Dev (1) stop			
TOTAL				

B

Figure 8–5. Camera reports.

Do you recall who assigned the roll number on your tracks? It was the mixer who recorded the numbers on the sound reports. Who assigns the roll number on your picture? The cameraman? Wrong! Your roll number is not on the camera reports. It first appears on the delivery slips and lab reports. Therefore, the picture roll numbers are assigned by the processing lab.

You are given three boxes of dailies. Double check the production numbers on all three delivery slips to be certain that they are *your* dailies and not another assistant's. When you return to the cutting room, take a few minutes to compare the camera reports with the script notes. Remember that the former lists *all* the takes and the latter, *only circled takes* so the circled takes on the *camera reports* (Figures 8–5 and 8–6) should correspond with the takes listed on the script notes.

CFI
CONSOLIDATED FILM INDUSTRIES
959 Seward Street
Hollywood, California 90038
(213) 462-3161

Company ROSALLEN
Address #8191
Production MURDER, ANYONE?
Date 10/4 Camera Report No. (TO ENSURE CONTINUITY OF DAILIES INDICATE)
Director ALLEN Cameraman BERGMAN
Magazine No. 3 Roll No. 4
Total Footage 1000 Emul. No. 5247-136

SCENE NO.	PRINT CIRCLED TAKES 1 5	2 6	3 7	4 8	DAY OR NIGHT REMARKS INT OR EXT
25I	(1)				Day Int
25J	1	2	3	(4)	
25K	(1)				
25L	1	(2)			

P-303
NG-642
W-55
T-1000
Print for Good Skin tones
Cool

1000
FORM NO. 10 (7-85) 750927

A

CFI
CONSOLIDATED FILM INDUSTRIES
959 Seward Street
Hollywood, California 90038
(213) 462-3161

Company ROSALLEN
Address #8191
Production MURDER, ANYONE?
Date 10/4 Camera Report No. (TO ENSURE CONTINUITY OF DAILIES INDICATE)
Director ALLEN Cameraman BERGMAN
Magazine No. 4 Roll No. 5
Total Footage 500 Emul. No. 5247-136

SCENE NO.	PRINT CIRCLED TAKES 1 5	2 6	3 7	4 8	DAY OR NIGHT REMARKS INT OR EXT
25L			(3)		Day Int
25M	(1)	(2)			Sil
25N	(1)				Sil
46	1	(2)			
30	(1)				

Print for Good Skin Tones
Cool
P-390
NG-48
W-62
T-500

500
FORM NO. 10 (7-85) 750928

B

Figure 8–6. Camera reports.

Of course, all the takes on the camera reports are listed in the order in which they were shot just as they are listed on the sound forms and script notes. Footages are given for the circled takes and the total listed as *good* (G) footage. The *not good* (NG) or uncircled takes are totaled, as well as unused or *wasted* (W) negative. The NG takes, or *outtakes,* are held in the lab or by a negative cutter assigned to your show as *B* negative for possible future use. The unused negative, also called *short ends,* is sometimes donated to schools to be used on student film projects.

When you open the daily cartons, you will find a *lab report* (Figures 8–7B, 8–8B, and 8–9B) for each picture roll. Compare each lab report with its accompanying

CUSTOMER COPY

CONSOLIDATED FILM INDUSTRIES

462-3161

A DIVISION OF REPUBLIC CORPORATION
959 SEWARD STREET
HOLLYWOOD, CALIFORNIA 90038

516731

ROLL CO. PROD. 8191 REQ. 466701

D1005-147 Rosallen Murder, Anyone?

35 MM

V.A. No Splice ☐	☐ 5234 Dupe	Re-order ☐ Paper to Paper ☐
Registered ☐	☐ 5243 Interpos.	ASSEMBLED BY RA / TESTED BY
Full Aperture ☒	☐ 5243 Dupe	CLEANED BY / PRINTED BY
16 MM Red. ☐	☒ 5381	DEVELOPED BY / INSPECTED BY
7381 ☐	☐ Direct Positive	
☐	☐ 5249 CRI B. & H. Perf. ☐	

Del. Print ☒ 1 - LITE ☐ TIMED

C.T. 25-23-25

Del. Neg.

NEG. FOOTAGE x NO. PRTS. = 550′ NEG. TOTAL + 23′ WASTE = 573′ TOTAL FOOTAGE

A

ROLL CO. PROD. 8191 REQ. 466701

D1005-147 Rosallen Murder, Anyone?

R	G	B		SCENE NO.	FEET	DAY	NIGHT	
25	23	25	1	39-2				
			2	39A-1				
			3	39B-1				
			4	24-1				
			5	25-4				
			6	25A-2				
			7	25B-3				
			8	25C-2				
			9	25D-3				
			10	25E-3				
			11					
			12					Neg. Footage
			13					550′
			14					

B

FORM 239 REV. 11-78

Figure 8–7. (A) Lab delivery slip and (B) corresponding lab report.

delivery slip (Figures 8–7A, 8–8A, and 8–9A), and you will find that all the information, including the production title and number, the individual roll numbers, and total footages enclosed in each case corresponds. For example, look at the delivery slip in Figure 8–7A and the lab report in Figure 8–7B. On both forms the production number is 8191; the title, "Murder, Anyone?"; and the roll number, 147.

Note that, though the negative (neg) footage is the same on both forms, a *waste* footage has been added only on the delivery slip. This additional footage is generally a 2 percent estimated charge for head and end leaders and any leader inserted between takes, all necessary for processing. (This waste footage accounts for the slight footage differences between the camera report and lab report.) The neg total is the estimated total footage of the scenes listed.

CUSTOMER COPY

CONSOLIDATED FILM INDUSTRIES

462-3161

A DIVISION OF REPUBLIC CORPORATION
959 SEWARD STREET
HOLLYWOOD, CALIFORNIA 90038

516732

ROLL CO. PROD. 8191 REQ. 466701

D1005-148 Rosallen Murder, Anyone?

35 MM

- V.A. No Splice ☐
- Registered ☐
- Full Aperture ☒
- 16 MM Red. ☐
- 7381 ☐
- ☐

- ☐ 5234 Dupe
- ☐ 5243 Interpos.
- ☐ 5243 Dupe
- ☒ 5381
- ☐ Direct Positive
- ☐ 5249 CRI

B. & H. Perf. ☐

Re-order ☐ Paper to Paper ☐

ASSEMBLED BY RA | TESTED BY | CLEANED BY | PRINTED BY | DEVELOPED BY | INSPECTED BY

A

Del. Print

☒ 1 - LITE ☐ TIMED

C.T. 25-23-25

Del. Neg.

NEG. FOOTAGE x NO. PRTS. = 830′ NEG. TOTAL + 30′ WASTE = 860′ TOTAL FOOTAGE

ROLL CO. PROD. 8191 REQ. 466701

D1005-148 Rosallen Murder, Anyone?

R	G	B		SCENE NO.	FEET	DAY	NIGHT	
25	23	25	1	25 F-4				
			2	25 G-1				
			3	25 H-2				
			4					
			5					
			6					
			7					
			8					
			9					
			10					
			11					
			12					Neg. Footage
			13					830′
			14					

B

FORM 239 REV. 11-78

Figure 8–8. (A) Lab delivery slip and (B) corresponding lab report.

On the delivery slip (Figures 8–7A, 8–8A, and 8–9A) *one-lite* is checked. One-lite will be explained in a moment. 5381 is the number given to the Eastman *positive* stock on which your dailies were printed. These numbers change from year to year to reflect changes in the stock.

The lab reports (Figures 8–7B, 8–8B, and 8–9B) list only circled, printed takes and are in shooting sequence. Items should correspond with those listed on the script notes and the circled takes on the camera reports. Any discrepancy between the takes you have received according to the lab report and those you should have received as indicated by the script notes should be attended to as soon as possible after you have finished syncing. If you are missing a picture take, call your lab contact person or the negative cutter or both depending on your company's policy. Order a color print from

CUSTOMER COPY

CONSOLIDATED FILM INDUSTRIES

462-3161

A DIVISION OF REPUBLIC CORPORATION
959 SEWARD STREET
HOLLYWOOD, CALIFORNIA 90038 516733

ROLL CO. PROD. 8191 REQ. 466701

D1005-149 Rosalleen Murder, Anyone?

35 MM

- V.A. No Splice ☐
- Registered ☐
- Full Aperture ☒
- 16 MM Red. ☐
- 7381 ☐
- ☐

- ☐ 5234 Dupe
- ☐ 5243 Interpos.
- ☐ 5243 Dupe
- ☒ 5381
- ☐ Direct Positive
- ☐ 5249 CRI B. & H. Perf. ☐

Re-order ☐ Paper to Paper ☐

ASSEMBLED BY RA | TESTED BY
CLEANED BY | PRINTED BY
DEVELOPED BY | INSPECTED BY

A

Del. Print ☒ 1 - LITE ☐ TIMED

C.T. 25-23-25

Del. Neg.

NEG. FOOTAGE x NO. PRTS. = 697′ NEG. TOTAL + 30′ WASTE = 727′ TOTAL FOOTAGE

ROLL CO. PROD. 8191 REQ. 466701

D1005-149 Rosalleen Murder, Anyone?

R	G	B		SCENE NO.	FEET	DAY	NIGHT	
25	23	25	1	25I-1				
			2	25J-4				
			3	25K-1				
			4	25L-3				
			5	25M-1				
			6	25M-2				
			7	25N-1				
			8	46-2				
			9	30-1				
			10					
			11					
			12					Neg. Footage
			13					697′
			14					

B

FORM 239 REV. 11-78

Figure 8–9. (A) Lab delivery slip and (B) corresponding lab report.

B negative (because it was not printed the first time), and furnish your production number and the number of the lab order that you should send directly as confirmation of your verbal order.

Should you receive an erroneously printed outtake with your dailies and your editor wants it included in the screening, order the sound track and incorporate it into your dailies.

The lab report has three columns marked *R, G,* and *B,* which stand for red, green and blue, the three colors necessary to produce a good color picture. To save time and expense the director of photography (DP), in consultation with the lab, has predetermined a *one-lite* constant setting for all the dailies unless otherwise specifically indicated. On the hypothetical dailies, as indicated on the lab reports the one-lite RGB has been set at 25–23–25. Any necessary retiming or adjustment of color from one scene to another will usually be done in the final processing to produce the best possible release print. Also, as noted on the camera reports, the DP has instructed the lab to *force one stop,* so that by providing one stop more light in the processing, a slightly brighter picture will result.

CALCULATING DAILY FOOTAGES AND SCREENING TIME

While you are checking your lab reports you should calculate your total daily footage and the screening time it represents. You will need these figures later to distribute your footage evenly over as few reels as practical and for scheduling your dailies. You should have surmised by now that the term *dailies* is used to mean either the picture and tracks to be synced or their screening. Examples of typical use are "Have you *built* (completed) your dailies?" or "When are your dailies?"

Having verified that the (neg) footages on the delivery slips agree with those on the lab reports, you may use either form for your footage addition. If you choose to use your delivery slips, be certain to use the neg totals, not the total footages that include the waste.

What is the screening time required for your dailies? First, if you compute the total footage, how do you estimate the screening time? And, secondly, if you should have a total running time, how can you arrive at the amount of footage to fill that time?

Since film running time is 90 feet per minute or 1½ feet per second, you can calculate either the time or the footage if you remember two formulas:

1. When you know the footage, figure the running time by:
   a. dividing the footage by 90 to arrive at the number of minutes or
   b. multiplying the footage by ⅔ to arrive at the number of seconds.
2. When you know the running time, figure the footage by:
   a. multiplying the minutes by 90 to arrive at the amount of footage or
   b. multiplying the seconds by 1½ to arrive at the amount of footage.

Let us test these two fomulae with a simple hypothetical situation. Assume, for example, that your picture dailies total 1800 feet. Running time can be calculated by formula 1:

a. 1800 feet divided by 90 feet equals 20 minutes or
b. 1800 feet multiplied by ⅔ equals 1200 seconds or 20 minutes.

In reverse, if you know you have 20 minutes of film, you can compute the footage using formula 2:

a. 20 minutes multiplied by 90 feet equals 1800 feet or
b. 1200 seconds multiplied by 1½ feet equals 1800 feet

Now that you have verified the mathematical formulae by using the above example, can you calculate the running time for *your* dailies, using either your delivery slips or lab reports for the necessary information? Try to do it before reading further.

Did you add the correct footages? If you used the delivery slips rather than the lab reports, did you use the *neg total* and not the *total footage?* Does your computation correspond with:

roll 147	550 feet
148	830
149	705
Total	2085

Did you arrive at *2085 feet* as the total footage of the scenes to be screened in your dailies? You should have figured the running time using formula 1:

a. 2085 feet divided by 90 feet equals 23 minutes with 15 feet remaining
   15 feet multiplied by ⅔ equals 10 seconds
   Total screening time equals 23 minutes and 10 seconds (23:10) or
b. 2085 feet multiplied by ⅔ equals 1390 seconds or 23:10.

Now do it in reverse. You have 23:10 screening time to fill. Calculate the amount of footage you would need using formula 2:

a. 23 minutes multiplied by 90 feet equals 2070 feet
   10 seconds multiplied by 1½ feet equals 15 feet
   Total footage required equals 2085 feet or
b. 1390 seconds multiplied by 1½ feet equals 2085 feet.

FOOTAGE CHART

There is an even easier, faster method than the above formulae and that is the *35mm footage chart* (Figure 8–10), copies of which both you and your editor should have in your cutting room within immediate vision. There are two sets of double columns: the two columns on the left show seconds and footage and the two columns on the right show minutes and footage. As you can see, you can find 1800 feet in the *footage* (fourth) column and in the *minute* column (column three) you can see the running time of 20 minutes. Do not confuse the time columns with the footage columns nor the minute and second columns. These are common errors.

The total footage on your dailies is 2085 feet. In the fourth column under *footage* go to the nearest footage below 2085, which is 2070. Opposite it in the *minutes* column is 23. But, subtracting 2070 from 2085, there is still 15 feet unaccounted for. In the second column under *footage* you will find 15 and opposite it, in the first column, is 10. You have arrived at the same running time, 23:10, as you did previously using the formulae.

SECONDS	FOOTAGE	MINUTES	FOOTAGE
1	1 1/2	1	90
2	3	2	180
3	4 1/2	3	270
4	6	4	360
5	7 1/2	5	450
6	9	6	540
7	10 1/2	7	630
8	12	8	720
9	13 1/2	9	810
10	15	10	900
11	16 1/2	11	990
12	18	12	1080
13	19 1/2	13	1170
14	21	14	1260
15	22 1/2	15	1350
16	24	16	1440
17	25 1/2	17	1530
18	27	18	1620
19	28 1/2	19	1710
20	30	20	1800
21	31 1/2	21	1890
22	33	22	1980
23	34 1/2	23	2070
24	36	24	2160
25	37 1/2	25	2250
26	39	26	2340
27	40 1/2	27	2430
28	42	28	2520
29	43 1/2	29	2610
30	45	30	2700
31	46 1/2	31	2790
32	48	32	2880
33	49 1/2	33	2970
34	51	34	3060
35	52 1/2	35	3150
36	54	36	3240
37	55 1/2	37	3330
38	57	38	3420
39	58 1/2	39	3510
40	60	40	3600
41	61 1/2	41	3690
42	63	42	3780
43	64 1/2	43	3870
44	66	44	3960
45	67 1/2	45	4050
46	69	46	4140
47	70 1/2	47	4230
48	72	48	4320
49	73 1/2	49	4410
50	75	50	4500
51	76 1/2	51	4590
52	78	52	4680
53	79 1/2	53	4770
54	81	54	4860
55	82 1/2	55	4950
56	84	56	5040
57	85 1/2	57	5130
58	87	58	5220
59	88 1/2	59	5310
60	90	60	5400

Figure 8–10. 35mm footage chart.

Footage		*Time*
2085		
−2070	=	23 minutes
15	=	10 seconds
Total Time	=	23:10

There are obviously numbers that do not appear on the footage chart. Assume that you are determining the running time of a total footage of 6690 feet, a footage figure that does not appear on the chart. Begin by selecting the footage figure on the chart closest to your total, in this case the last figure on the chart in the fourth column, 5400 feet, which is 60 minutes or one hour of running time.

Subtract 5400 from your original total of 6690 feet and you are now left with 1290 feet. The footage number on the chart prior to this remaining footage is 1260, which represents 14 minutes of running time. You thus have, so far, 1:14:00 (time is stated as hours:minutes:seconds). Subtracting 1260 from 1290, you now have 30 feet of film unaccounted for. On the chart, using columns one and two, you find this equals 20 seconds of running time (perfectly reflecting the ⅔ conversion ratio). Adding the 1:00:00, 14:00 and :20 figures gives you a total approximate running time of 1:14:20.

Footage		*Time*
6690		
−5400	=	1:00:00
1290		
−1260	=	14:00
30	=	:20
		1:14:20

It is important to remember that all footage or time figures should be approximated, whether using the mathematical formulae or working directly with the chart. For example, in the calculation below, the closest footage figure to 56 is 55½, representing 37 seconds, so this figure should be used.

Footage		*Time*
3206		
−3150	=	35:00
56	=	37*
	=	35:37

*Note that this figure is approximated to the nearest second.

This procedure can be used to calculate either running time from footage or footage from running time. You simply reverse the procedure when you require the footage for a certain amount of time. What are the footages for the following: (a) 2:27:04 and (b) 7:47?

(a)

Time		*Footage*
2:27:04		
2:00:00	=	10,800
27:00	=	2,430
:04	=	6
		13,236 total footage

(b)

Time		*Footage*
7:47		
7:00	=	630
:47	=	70½*
		700 + 08 total footage

* ½-foot = 8 frames

You will soon learn that measuring footage and computing running time is a vital part of your responsibilities as an assistant film editor. While it is not imperative to know the running time to the absolute second for your daily screenings, in little more than a week you will have to provide measurements of the cut or edited picture to the footage plus frames and compute the running time to the exact second for the first cut or screening for your editor and director.

But you have much more to learn before then. You are now ready to sync your first day of dailies.

CHAPTER 9

SYNCING DAILIES

The hypothetical picture dailies you brought to the cutting room from shipping and receiving were on cores and were generally tails out and cell up. You should rewind them onto reels so they are heads and emulsion up. You have three rolls of film in three cartons and, as you remove each roll from its box, keep the lab reports attached to their respective film so that you have a reference for what scenes are in which rolls.

First, let us assume that your editor has approved the assembly of the dailies in shooting sequence. Before you rewind your picture to heads, check whether there are any scenes that have to be shifted for similar scene sequence, as explained in Chapter 8, because you will want to splice out such scenes as you are rewinding.

Take one of the rolls and place it on a (male) flange so that the (tail) end of the film will come off the *bottom* of the roll. Place the flange on the left rewind and draw the film left to right across the bench attaching it to an empty reel on the right rewind so that the film will wind clockwise on the reel. In other words, besides winding the film to heads, you will be winding the film *under and over* so that the film will go from being cell up to being emulsion up.

Should a roll be delivered to you differently, you must adjust the procedure accordingly. Should it be emulsion up, tails out, you want to retain the emulsion side on the up-side by winding the roll off the *top,* instead of the bottom, so you will wind the film *over and over* to make it heads up.

It saves time if you rewind all three of your rolls at the same time. You can easily do it if you have three split reels (Figure 4–8) that securely hold the film on cores. You may not have any split reels available however. You might still be able to do it by placing one roll on the single flange as described above followed by the other two rolls with only their cores embracing the rewind shaft. As you rewind left to right onto three reels now, the last two rolls on the left rewind will bounce around considerably requiring that you maintain firm control with your left hand.

After you have rewound all three picture reels of your hypothetical dailies, take the first reel beginning with scene 39–2 and the corresponding track, and place them on your left rewind, with picture closest to you. Put two empty reels, for takeup, on your right rewind and attach the picture and track daily leaders marked *Reel 1 Code A–1000* that you prepared the previous day (see Chapter 6). Place a clamp on the right rewind to hold the two take-up reels together.

Place the picture start mark in your first gang *in frame* (see Chapter 3) and align the track start directly opposite in the second gang. Rewind to the end of the leaders, but before you allow the leader ends to run through the synchronizer splice the end of the picture leader to the head of your first slate, 39–2, always making sure you are splicing in frame (on the frame line). Now you can let the track leader run through free of the synchronizer and unspliced.

Remember that syncing dailies is simply keeping picture and track *even,* syncing each take individually by using the visual and audible clapsticks as principal references. Also remember that you are matching track to picture, *not* picture to track. The syncing of the first take of a daily reel is slightly different than the syncing of takes on the rest of the reel. That difference will be noted in the following six steps for syncing:

1. Choose a frame line a few frames after the last frame of a take. This spot will either be in the slate area of the next take you have to sync or in some leader between the takes. Position this spot in the center of your synchronizer and mark it with your white grease pencil on the picture and on precisely the same point on the track. (Always use your synchronizer for aligning.) Make both marks on the sides of the film farthest from you, and bring the white line halfway down the frame line. These marks will serve as a reference point, henceforth referred to as your *personal sync marks (PSM).* (For the first take of each reel, consider the tail ends of your daily leaders PSMs.)
2. Throughout this syncing procedure use the hand roller on the synchronizer rather than the rewind. Roll the film to the right where the splicer is. (If you are syncing the first take of a reel, the tail of the track leader will thus run out of the synchronizer.) Turn the track either toward you or away from you so that it is cell side up and place it on the butt splicer (Figure 4–12). Cut the track at the PSM. Do not cut the picture. Free the left part of the track from the splicer but leave the right portion in place for future splicing.
3. Now roll down and find the precise frame on the picture where the upper and lower sticks of the clapboard initially touch. You can usually find this by visual inspection (called *eye-balling*), with a magnifying glass, or as a last resort by using a moviola. As you do this, make certain the scene or slate identification corresponds with that on the lab report. Mark the initial frame with an *X* and write the identification next to it, which in your hypothetical dailies would be 39–2 (Figure 9–1). Lock frame *X* into the center of your synchronizer.
4. Release the track from the sync machine. (If this is the first take of a reel, the track has not yet been engaged in the synchronizer.) This is the only time you

Figure 9–1. Clapsticks.

release the track. If possible, you never release the picture so that you will eventually be able to record code numbers while you are syncing as will be explained later in this chapter. Find the scene or slate on the track you earlier popped and line up the pop mark with the center of the picture frame marked *X*. Lock the track back into the synchronizer being careful to keep the pop mark symmetrical with the picture frame marked *X*. With experience you will learn to adjust the track pop a sprocket or two according to how you estimate the picture clapstick first touched.

5. Return to your PSM on the picture, and mark a new corresponding PSM on your track. Occasionally, you may have to add leader to the track to return to this position. Remember that any kind of film that you use for leader must be cell, or shiny side, up so you will splice it on the emulsion side. Since your daily track is the reverse, you will splice emulsion to cell.
6. Turn the track in the same direction you turned it in the second step above. Otherwise you will end up with a twisted track. Cut it on the new PSM and splice it to the right-hand segment of track that you left in the butt splicer. (If this is the first take of a reel, you will splice to the tail end of the track leader.)

With a few exceptions such as end slates and lip syncing this is all you have to do on each take to keep picture and track even and in sync with the clapsticks. Do you wonder why, to begin with, there is any discrepancy between picture and track? After all, was not each take in sync when it was shot? Certainly they were, but remember that when the take was shot "*marker*" was not called until the mixer had the sound up to speed. So there was extra track there. In addition, when the lab removed the outtakes from the negative rolls, rarely was any leader added between the printed takes. But when the sound department transferred only the circled takes, some additional blank sound track invariably resulted between each take as the transfer person made certain everything required was recorded. This should explain to you why you usually receive more track than film in your dailies and why each take must be individually synced.

The more takes you have on a reel, the longer it takes to complete that reel. How many takes should you put on one reel? It does not matter, but you should try to consolidate your dailies so that there are as few reels as possible to be transported to the projection booth and the coding room. At the same time, if it is feasible, it is preferable to make the amount of footage on each reel even rather than overload one reel. Overloading creates a danger that if the film is not tightly wound it will flop over the side of the reel in projection or coding. This may result in injuring the film.

How do you distribute the total daily footage evenly over an appropriate number of reels? You previously determined the total footage by adding up the footages listed on the lab reports, which should have coincided with those on the delivery slips. Picture rolls you receive from the lab may be as little as a couple of hundred feet or as much as 1000 feet. Let us imagine that you have received five or six rolls of various footages from the lab and they add up to 1800 feet. About 900 feet on each of two reels would make an ideal distribution—of picture and track separately of course. (When the singular term is used in reference to *a* reel of dailies or edited picture such as "Is that reel ready?" or "Give me reel 2", *both* picture and track are implied.)

Suppose that toward the end of reel 1 you see you have about 750 feet of film and, by checking the footage of the next take on the camera report, you learn it is over 300 feet making a total of 1050 feet. You know you can not have that much footage on a 1000-foot reel, and you certainly know that you *cannot* begin a take at the end of a reel and continue it onto the next reel. You will have to end reel 1 with only 750 feet and begin the next reel with the 300-foot take. In this particular case you will be forced to *build,* or assemble, three reels of dailies. In any case, take a few minutes to preplan which takes are going to end or begin your reels.

Once you have arrived at the end of a reel, remember to mark *end syncs* on picture and track for the apprentice coder as was previously discussed in Chapter 3. In addition, add about six feet of end leader to each reel. If actual leader is scarce or you wish to economize, for picture use lab leader from the daily rolls emulsion up and for sound use excess track cell up. Identification at the tail end of the end leaders for picture (in black pen) and for track (in red pen) should include

1. *End Picture* and/or *End Track*
2. Beginning code number
3. Reel number
4. Production title and production number

When you have finished syncing all the reels, rewind them to their heads, and they will be ready for screening. If you wish to double check your work, as you finish each reel rewind picture and track *in the synchronizer*, initially lining up those end syncs, and as you rewind, spot-check the clapsticks and finally the start marks to be sure they are still in sync.

If, while you are double checking, the clapsticks on one of the takes are out of sync, detach the tape on the track at the PSM and resync this take correctly. This next step is very important. Although you have corrected this take, the remainder of the reel has been adversely affected. You must rewind the film head to tails and *resync the next take.* If, in the corrected take, you have to *eliminate* excess track, save that piece of excess track and add it cell up as leader to the end of the corrected take. The following take, as well as the remainder of the reel, should still be in sync. If you have to *add* track, then you will have to *eliminate* the same amount of track prior to the slate of the following take. With both takes now in sync the remainder of the reel should also be in sync.

But you still have not completed double checking the takes *preceding* the corrected takes. You must reverse the winding and resume winding tails to head and check for sync up to and including the start marks. Only then can you be assured everything is in sync.

SYNCING END SLATES

Once you become adept at the six steps involved in syncing dailies you will generally be able to prepare for any screening schedule despite a possible delay in lab delivery or unforeseen difficulties with your dailies. One such variance that might occur is end slates, which we described in Chapter 8. An end slate is at the end of a take instead of at the beginning and is positioned upside down so that it will not be confused with the beginning slate of the following take. An end slate can require you to spend extra time syncing, particularly if it is a long take.

Accommodating an end slate requires only minor changes in syncing procedure. In the second step, instead of leaving the right track in the splicer after you have cut the track, paper clip the two ends of track together so that you can use your rewind to wind down to the end of the take for the third and fourth steps. Why then cut the track? The paper-clipped track will help you find your personal sync mark faster when you wind back through the synchronizer to the head of the take for the fifth step. Make a new PSM on the track, remove the paper clip, replace the right track onto the splicer, and complete the sixth step.

THE MOVIOLA

An even greater delay results when picture clapstick and track pop are questionable or nonexistent. This means you have to establish the synchronization by *lip sync*. This requires the use of a *moviola*. Assistants on features, miniseries, and TV movies will usually have their own moviolas, but occasionally, when a company decides to economize in this area, it is the assistant on an episodic series who has to do without.

In such a situation an astute post-production supervisor will try to provide a community moviola in an otherwise unused cutting room within easy reach of most of the assistants. The other alternative is for you to use your editor's or a neighbor's machine, with permission of course. Whoever's equipment you use always be careful, no matter how pressured you might be for time, to leave the film, bench equipment, and moviola exactly as you found them.

Moviola is both a registered trade name and a generic term for upright, motor-driven film viewers used in editing. The Moviola (the original machine) has been the perennial editing machine since the early 1920s, beginning as a bench viewer and with the advent of sound in the 1930s becoming a self-standing, upright machine. Its European counterpart is a horizontal, or flatbed, machine that began to be used increasingly by some Hollywood editors during the last decade. Irrespective of the flatbed and despite impressive technological advances in videotape editing, the moviola still perseveres as the major equipment for most Guild editors. Because of this popularity and its exclusive use by film editors on episodic TV series, I will assume that you will be using a standard, upright moviola.

It is a hardy machine, requiring little maintenance. However, a moviola can be a frightening piece of equipment for the novice operator, churning the film through its sprocketed teeth, seemingly about to tear it to shreds at any moment, chattering and clattering as it roars along. But in comparison to the computerized tape machines it is simple to operate. Once you learn to run and control it, like driving a car, it will become automatic and will become your ally.

Let us assume that you are a fortunate assistant who has been assigned your very own moviola. It is a *cutter moviola* used by picture editors as opposed to a *take-up moviola* (Figure 11–5) used by sound and music editors. The major difference between the two machines is that the cutter moviola has a large cloth bag attached to its rear for the film to drop into after projection whereas the take-up moviola has arms attached to the rear of the machine that are used to hold reels so that picture and track can be mechanically wound up as on a projector. The occasional use of a take-up moviola by you or your editor is discussed in Chapter 11.

Study the cutter moviola as illustrated in Figure 9–2. Before you begin running a moviola, you should know about its parts.

Make certain the power cable is plugged into an electrical outlet and into the moviola receptacle labeled 1. The master switch that turns on the machine is 2. The forward/reverse switch for the sound is 3. As will be explained, this switch is usually set on *forward.* The sound head motor switch that will run the machine automatically at constant speed is 4. You will rarely switch it on, however, because you have safer control using the foot pedals. It is sometimes used when running film for a producer or director. The lever that, when pushed down, will open the sound gate is 5. The

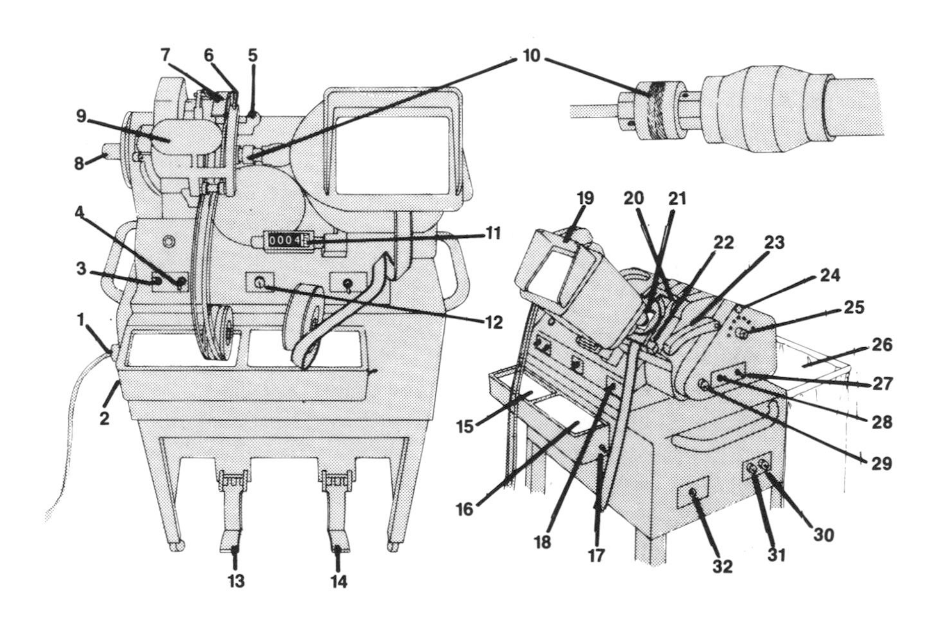

Figure 9–2. Moviola and its parts.

sound gate is 6. When the sprockets of the track are threaded onto the pins, the gate is brought down to secure the track. The sound head, 7, is attached to a small bar that permits it to be raised or lowered for contact with the track. It can be slid into any one of three positions. Should you have occasion to run *full coat* with sound on three channels containing, for example, sound effects on the left stripe, music in the middle, and dialogue on the right, the sound head can be easily moved to either position. Its most customary position is on the right side so that it can be placed on the sound track of the single-stripe mag with which you and your editor will be principally working. The starting point you select on the track, whether it be a particular code number or the pop, should be positioned directly beneath the center of the sound head. The *moviola flange* used to wind up a roll of film is 8. The sound system for optical tracks that you will rarely use is 9. If you recall, as I discussed in Chapter 2, optical tracks are on composite prints. Optical sound operates with a prefocused *exciter lamp* and is situated on the left side of the track gate so that it will contact the optical track on the left side of the film.

A drive shaft connector that, if it is in the customary left position, enables picture and track to run concurrently and in synchronization is 10. This is termed as being *in interlock.* If the connector is released from this position by pulling it to the right, the drive shafts are *out of interlock* or disconnected so that picture and track will run separately. The footage counter is 11. The on/off sound knob with volume control is 12. The left foot pedal, 13, is called the *constant* pedal because it will cause the machine to run at constant speed of 24 frames per second. The *variable* pedal, 14, can make the machine run as slowly as a frame at a time or as fast as 30 frames per second depending on the pressure exerted by your foot.

The concave cover for a small compartment beneath the machine is 15. Small items such as a scissors, a stylus, or a spare projection bulb are stashed in the compartment below the cover and paper clips are usually kept in the cover itself. The glass-covered well light that is helpful in reading key numbers is 16. The switch for the well light is 17. The forward/reverse switch for the picture is 18 and is usually set at *reverse.* The picture viewer is 19. When you desire it, the viewer can simply be lifted away to one side. A framing lever to adjust the picture in frame is 20, and 21, next to the projection lens, is a mechanism to adjust the picture focus. Number 22 is a button that releases the gate so that the picture film might be threaded onto the sprockets. Then the gate can be locked into place by pushing it closed. The start mark or code number is placed in the gate aperture. The brake that will instantly stop the machine even though you may still have your foot on a pedal is 23. Number 24 is a screw attachment that secures the right side panel of the moviola. When it is unscrewed, the panel is ajar so that you can replace a projector bulb. The rheostat knob for the variable speed (picture) motor is 25. The cotton moviola bag into which the film falls after projection is 26. The picture motor on/off switch is 27; the picture light off/on switch, 28; the rheostat knob for the picture light, 29; the sound tone knob, 30; and the mag/optical knob selector, 31. This last knob is usually set for magnetic sound, but, should you have to run optical track, you must reset it accordingly. Finally, 32 is a jack for headphones that you may need when both you and your editor are running moviolas in the same room.

LIP SYNCING

During the syncing of the dailies you may have to use a moviola to lip sync a take for a number of reasons: some questionable interference with the track pop, multiple pops or picture clapsticks, or having to match up picture and track with misnumbered takes and illegible or missing slates or no announcements.

Somewhere in the middle of a reel of dailies you come to a scene that, for one of the above reasons, you must lip sync. At the bench on the left rewind are the unsynced, or source, reels of picture and track and on the right rewind are the take-up reels containing the completed, synced scenes. Execute the first two steps of syncing: mark the PSM then cut the track at the PSM.

You must feed picture and track off of the left rewind into the moviola. You will have, hopefully, a swivel base on that rewind so you can angle the reels toward the machine. The left track can now be freed of the synchronizer while the picture should remain intact, retaining any counter footage you may have set. Release the clamp on the right rewind temporarily so the take-up track reel will remain stationary and not be affected by the movement of the corresponding picture reel. Pull the picture back through the left side of the sychronizer, providing about 10 feet of slack ahead of the slate for ease in reaching the moviola and for directing the film into the moviola bag. The track, cut at the PSM and free of the sync machine, can then be lined up with the moviola. Since you are having a problem with either the pop or the clapstick or both, line them up where you *think* they might be.

Wear a glove on your left hand to prevent nicking the film or cutting a finger and gently but firmly hold both picture and track so as to guide them and relieve the tension while you gently hold the picture below the picture gate in your right hand to guide the film and to give you instant access to the brake. Use your right hand to adjust the focus or framing, if necessary. Set the left switch on forward and your right switch on reverse. When you press the left foot pedal, the film will run at a *constant* speed of 90 feet per minute. Depending on how hard you press the right foot pedal, the film will run faster or slower, at a *variable* speed. Even though both switches are reversible in ordinary running get in the habit of using the left foot pedal for forward and the right one for reverse. Lip syncing, however, demands more than ordinary running.

You can now run the film. Make sure it is dropping into the bag at the rear of the moviola and not flopping over the side onto the floor. Run as little film as possible. This is not an opportunity to savor the entire scene as you still are not certain how many other takes have to be lip synced, and you still may not know whether you have an early screening. You must work fast, or you may not be ready in time.

You need to select either a very distinct sound on the track or the corresponding picture, whichever you can first identify. In an action take it might be a shot of a pistol or rifle, the slam of a car door, or the sound of an object hitting the ground. In a dialogue scene, where you can see the performer's lips, listen for a word with a consonant that causes the speaker to part his or her lips visibly, a word such as *money, people,* or *bottle.*

At the sound or related picture, whichever comes first, mark the first sprocket of sound or the exact frame of picture with a white grease pencil. While retaining this first mark where you have it in the machine, you must now find the corresponding picture or sound. In order to do this you must be able to run either one individually so you must shift the moviola to *out of interlock.*

The *interlock* (Figure 9–2) is a coupling bolt connecting the drive shafts between sound and picture. When it is positioned in the usual manner, at the extreme left, the moviola is *in interlock,* meaning that sound and picture will run concurrently. When the coupling is moved to the extreme right, it is then *out of interlock,* and either picture or track may be run individually in forward or reverse directions. You must use the right pedal for picture, limiting it to variable speed, and the left, or constant, pedal for the track.

As soon as you find the related sync, mark it and line it up in the moviola with the first mark. Shift the coupling to the left to put the machine back into interlock, and run the machine long enough to assure yourself that the take is in sync. You may have to make a slight adjustment, or if you are still uncertain and can find another distinct sound, sync that one up as a double check. You should mark this second set of syncs with a *2* or some other mark so as not to confuse it with the first set.

There are two methods by which you can return the film to the bench and resume syncing. First, remove the picture and, holding it high with your left hand so you can guide it and prevent any possible entanglements as the film twists from the bag, with your right hand at the right rewind wind the picture until your selected sync mark reaches the center of the synchronizer. Take the track from the moviola and line up the corresponding sync. In the second method you reverse the film in the moviola back to the slate, mark new corresponding sync marks on both picture and track, and use these syncs to line up in the synchronizer. The advantage of this second method is that, returning to the slate in the moviola, sometimes a questionable pop or clapstick can be resolved or re-enforced by your lip sync.

Before having to lip sync, you complete the first two steps of syncing the take. Lip syncing has completed the third and fourth steps. Now you can continue on with the last two steps. Then, as you wind down to the next take, remember to wipe off the lip sync marks on the picture since you do not want them to be seen in the screening. If you wish to retain those marks for possible future reference, mark the frame on the *edge* of the film.

CODE SHEETS

If you have time before the dailies are screened, make up code sheets, which are your personal record of the film received and of its assembly. Basic information that should be noted on the sheets includes production number and title, date received, date shot, scene and take (or shot) number, code numbers, reel numbers, negative and sound roll numbers, and other remarks.

Reel 1 (Figure 9–3) begins with scene 39, take 2, and goes through 25F–4 in shooting sequence. Figure 9–4 continues with reels 2 and 3. Place the first reel of picture only in the synchronizer at the start mark and set the footage counter at 0000. It is not necessary to set the first digit, the 1000 or 2000 number. You know that, beginning at this mark, both picture and track will be coded A–1000 as specified on the reel

Columbia Pictures Television

EDITORIAL DAILY FILM CODE RECORD

PROD # 8191 | TITLE: Murder, Anyone? | DATE: 10/5/

USE BLACK PEN OR PENCIL ONLY - MAKE ONE CARBON COPY - HOLD COPY FOR SOUND FX

DATE SHOT	SCENE NUMBER	SHOT NUMBER	CODE NUMBER	NEGATIVE KEY AND NOTES	NEG ROLL CAN NO.	SOUND ROLL NO.
10/4	39	2	REEL (1) A-1000		147	14
↓	39A	1	35		↓	↓
	39B	1	54			
	24	1	77			
	25	4	141			
	25A	2	217			
	25B	3	293			
	25C	2	366			
	25D	3	436			
	25E	3	515		↓	
↓	25F	4	552		148	↓
			1822			
10/4	WT 25-1			Sc. 25 Marshall's dial.		14

LIST ALL WILD TRACKS - DELIVER WILD LINES TO LOOPS AND WILD FX TO SOUND FX, WHEN THE WORK PRINT IS TURNED OVER TO SOUND FX & MUSIC

ASSISTANT EDITOR Erik Rose EXT NO. 1427 PAGE 1 OF 2

Figure 9–3. Code sheet, reel 1.

1 daily leaders; therefore, the last three digits on the footage counter will now correspond with those on the code numbers to be imprinted. As you roll the reel, write the code number as indicated by your counter as it appears at the head of each take as well as the pertinent information listed in the previous paragraph. Also, WT-1 for scene 25 should be noted separately near the bottom of the page. (See Chapter 10 regarding the coding of wild tracks.)

Columbia Pictures Television

EDITORIAL DAILY FILM CODE RECORD

PROD # 8191 TITLE: Murder, Anyone? DATE: 10/5/

USE BLACK PEN OR PENCIL ONLY - MAKE ONE CARBON COPY - HOLD COPY FOR SOUND FX

DATE SHOT	SCENE NUMBER	SHOT NUMBER	CODE NUMBER	NEGATIVE KEY AND NOTES	NEG ROLL CAN NO.	SOUND ROLL NO.
			REEL (2)			
10/4	25G	1	A-2000		148	14
↓	25H	2 E.S.	282		↓	15
↓	25I	1	556		149	↓
↓	25J	4	584		↓	↓
↓	25K	1	849		↓	↓
			2861			
			REEL (3)			
10/4	25L	3	A-3000		149	15
↓	25M	1 MOS	230		↓	
↓	↓	2 MOS	269		↓	
↓	25N	1 MOS	285	292-301	↓	
↓	46	2 NO ANN	312		↓	15
↓	30	1	358		↓	↓
			3396			

LIST ALL WILD TRACKS - DELIVER WILD LINES TO LOOPS AND WILD FX TO SOUND FX, WHEN THE WORK PRINT IS TURNED OVER TO SOUND FX & MUSIC

ASSISTANT EDITOR Erik Rose EXT NO. 1427 PAGE 2 OF 2

Figure 9–4. Code sheet, reels 2 and 3.

You have two choices that differ from the code sheets illustrated. On reel 1, for example, A1000 is not the beginning of the first take. The reel was coded A1000 from the start mark and there is four feet of leader before the first take so you may want to indicate the exact beginning code number of the take rather than the reel, which would probably be A1005. Your second choice regards the manner of recording the code numbers. In the illustrations, except for the first number and the end number of the last take, only the last two or three digits are indicated. You may want to use full numbers for each take.

The footages on your code sheets will not correlate exactly with those on the camera reports. The reason is that the lab, as previously mentioned, may add varied amounts of leader between the printed takes. Sometimes the first frame of a slate will come immediately following the last frame of the previous take. In the code sheets illustrated one foot of hypothetical leader has been selected to represent these differences between each take.

Some companies require that a *focus chart* be placed on each picture reel between the tail end of the daily leader and the first take. This chart consists of a diagram of vertical and horizontal lines that aid the projectionist in establishing the ideal focus before the first take so that the screening is not disturbed. The chart film should be about ten feet in length with, of course, corresponding leader footage on the track to keep it in sync with the picture. If it is required, you should add the focus chart when you build the daily leaders. With this addition, in the above example the exact code number of the first take on reel 1 would then be A–1015.

In addition to the wild track you listed on the reel 1 page, the end slate (ES) or end marker (EM) on reel 2, and the silent (*MOS*) and no announcement (*no ann*) notations on reel 3 other remarks that should be included on your code sheets, when they are applicable, are picture or track missing, multiple cameras, second slate, lip sync, misslate or misannouncement, outtake, reprint, lab defect report, and takes with camera stops. All these remarks are for editorial reference, but some of them—any missing material, outtakes, reprints or defect reports—should be indicated and explained to those viewing the dailies, and you should mention them to your editor *before* dailies are screened.

An *outtake* in this case would be a take that had been mistakenly circled or printed by the lab. Ask your editor whether these outtakes should be included in the dailies. If so, you will have to order the track. If not, do not discard them. File them in a trim box marked *Outtakes;* they may be useful later. *Reprint coming* noted on your lab report means there was some problem with the print, usually in its color, and a corrected print will be sent later in the day. It may be delivered in time to replace the temporary reprint before screening or coding. If not, you will have to determine whether you wish to wait for it before coding the concerned reel or code the reel and code the take separately.

A *lab defect report* will come to you in various ways. It may be sent to the post-production supervisor's office, directly to you, or by phone. Such a report will indicate the scene and take number and the frame or frames by key number of one of the daily takes that has some sort of negative defect.

As you become more experienced in syncing dailies, you will be able to make up the code sheets in one operation while you are syncing. To accomplish this you must not lose the footage, retaining an accurate count at all times on the picture. You have now learned how to leave the picture in the synchronizer while lip syncing without disturbing the footage. But what happens when you must lip sync on a moviola in another room? In that case pull the picture back so that you temporarily cut it at your PSM *left* of the synchronizer either by pulling your picture over to the butt splicer, which is at the right of the synchronizer, or by simply moving the butt splicer to the left of the machine. Lock the synchronizer so the film will not accidentally move and the footage will hold fast. You are now free to remove picture and track reels from the left rewind and take them elsewhere. When you return, splice the picture back together. So long as you do the splicing on the left side of the synchronizer you will be able to maintain the footage. Also, keep this procedure in mind when you have to add picture to a reel such as when you are assembling dailies in numerical scene sequence.

If you are unable to make up your code sheets before dailies are screened, you must bring to the screening a list of your assembled takes by reel numbers either on your incomplete code sheets or on a sheet of paper so that you will be able to point out those remarks I previously mentioned and to note any comments made in the projection room pertaining to any of the takes.

EXCEPTIONS

The six-step procedure for syncing dailies that has been described is, in my opinion, the simplest, fastest, and most nearly error-proof method. It is designed so that only two gangs of the sync machine are used. Therefore, the method of operation remains the same whether you are given a two-gang or three-gang synchronizer. Rarely will you splice off and lose any track information preceding the pop; you will usually be able to hear the track announcement in dailies should you have to recheck that area. This is by no means the only method of doing the job, however; there are several others and probably some I have never heard of.

A word of caution: If you are in a subsidiary position and the person with whom you are working insists you do dailies her way, be flexible and comply. You will keep your job and establish a reputation for being cooperative. And you may find you prefer that method to the one I have suggested.

KEY NUMBERS ON CODE SHEETS

Some editors still insist that their assistants include beginning and end key numbers for each take on their code sheets. Generally, you would use the last key number on the slate for the beginning number and the last number at the end of the take for the end one. Though some editors will undoubtedly disagree, I do not believe that the time and effort recording key numbers requires is warranted for the rare occasions when this information is useful, particularly on episodic television where upright moviolas are used and individual trim rolls are filed, making it easy to check the numbers. (See Chapter 11 for further explanation.)

Requiring key numbers to be recorded goes back about 30 years to a time before individual take rolls were used as a method of filing trims. Trims of multiple takes were then consolidated into huge *cartwheels* making it difficult to find a particular take for the purpose of checking a key number. Requesting key numbers on code sheets, however, if the editor is using a flatbed rather than a moviola, is not unreasonable. In a flatbed situation trims are filed on *kem rolls* (see Chapter 10) and, while it is not as difficult as working with the cartwheels of 30 years ago, it does take some time to roll down to a particular take.

THE FLATBED

The most commonly used flatbed, or horizontal, editing machines in Hollywood are the Arriflex®, Kem®, Moviola®, and Steenbeck®. Although Moviola does make a horizontal machine, its name as a generic term refers to the upright while *kem* is a loosely-used generic term for the flatbed and for the daily or work rolls prepared for those machines.

Moviolas monopolize the editing of episodic television so the use of the horizontal machines is of no immediate concern for you in the hypothetical situation in which you have been placed. Flatbeds, however, are preferred by some feature editors and some editors who work on television specials and miniseries so it behooves you to learn as much as you can about their use and requirements when you have the opportunity.

Why are horizontal machines not used on episodic series? First, most series editors prefer moviolas simply because their basic experience has been with upright machines. More decisive reasons are purely economic. The flatbeds are large, bulky tables that occupy more space than is usually available. They cost or rent for about three times as much as the uprights, and an editor usually requires at least two assistants when using one. Television series do not budget for such extra editorial expenditures.

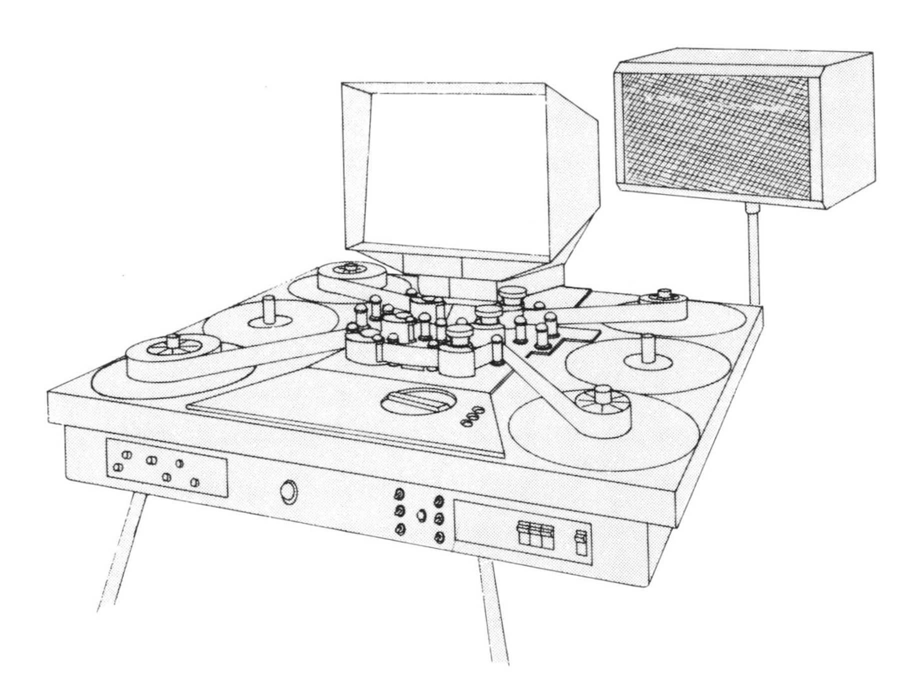

Figure 9–5. Horizontal editing machine (flatbed).

Flatbeds (Figure 9–5) come in various sizes, but all basically have flange-like metal plates on the left side of the *table* that hold picture and track rolls on cores. The film rolls off through picture and sound mechanisms in the center of the machine onto cores on plates to the right. (Instead of cores, the plates may be removed and reels used, but this may cause mechanical problems on some machines and is not often done.) Depending upon the sophistication and cost of the machine, it has from one to several screens for viewing.

If you do work on a project with a flatbed, its use should not affect the assembly of your dailies but the screening may be affected. Besides a perceptible savings in projection expense those viewing screens on the flatbeds provide producers and directors with the option of seeing dailies in the cutting room instead of being committed to a projection room schedule. Unfortunately, this also provides them with the option of arbitrarily postponing the dailies that often results in a delay in further editing.

SILENT TAKES

MOS takes are *midout sound* takes. (That expression or *mike off screen* is often offered as an explanation for the abbreviation.) In dailies these takes must have corresponding leader on the sound track to maintain sync, but many assistants fudge on the shooting of numerical scene sequence by placing their MOS takes at the *ends* of reels, so that they don't need to add corresponding track leader and simply end the track reel with the last sound take.

Aside from possible complaints from your superiors for not having those takes in proper sequence, you may have created another problem: The track end sync will still be at the end of the last take on the track aligned with the picture end sync just ahead of the first MOS take. While the end sync on the track would be easy to find, the apprentice coding the film will have neither the time nor the inclination to look for and line up the picture end sync that will be some distance from the end of the reel depending on how many silent takes you have. Therefore, any coding error that might be made will not be discovered until later when you break down your dailies thus resulting in a delay in corrective coding.

CAMERA STOPS

Sometimes there will be a take that begins with an ordinary slate and clapsticks, but then there will be a number of camera stops, each time with the action beginning again without benefit of either slate or clapsticks. Each of these takes within the one prevailing take may have the same angle and similar action or they may be somewhat different in both angle and action, but they will be identified under one scene and take number on the camera and lab reports. Subsequently, on the code sheet you will have to list the beginning code number at each camera stop (see 25N–1 in reel 3, Figure 9–4).

These takes with camera stops are usually MOS and can be taken in stride, but when they do have sound, your work is significantly increased. Even though you can easily sync the beginning of the take with the clapsticks, you may have to lip sync at each of the camera stops.

THE "DETECTIVE" ASSISTANT

It should be obvious how important it is for script notes, sound reports, and camera reports to be *in sync*. It makes the syncing of dailies uncomplicated and routine particularly if there are clear and distinct clapsticks and pops.

Lip syncing will interrupt that routine and can be a time-consuming process. But the real difficulty arises when the sound that supposedly corresponds with a picture simply does not match and you must become a detective and search for the right track.

How can this discrepancy occur? It may happen when the crew becomes careless and a misslate goes by unnoticed. It often happens on a location shoot where technical problems or weather conditions diminish efficiency, and particularly when a location crew, for whatever reason, is without a script supervisor.

When tracks are misnumbered, finding a track that will sync up with your picture can be exasperating. Back up to your preceding track and make certain it is the correct sound for the preceding picture. If it is, you can assume that the error in numbering began after that take with one of the outtakes prior to your printed take. Order the track outtakes and try each one until you find the right one. If that fails, then try succeeding takes. The correct track has to be there somewhere! You will find it by trial and error—and by being a good detective.

CHAPTER 10

DAILIES FINALIZED AND PREPARATION FOR EDITING

SCHEDULING DAILY SCREENINGS

Sometime in the morning, probably during syncing, you will receive a phone call from the person in charge of scheduling all screenings in the various projection rooms and theatres. This person may be a secretary or assistant to the head projectionist or someone in the post-production supervisor's office. She will want to know how much time you will need to screen your dailies. You should have an answer ready when you receive the phone call, because, when you added up the footages on the lab reports before beginning to sync, you should also have calculated the running time.

When she first calls, or at any rate before you have completed syncing, the schedule person will let you know when and where your dailies are scheduled. The screening may be later that morning or sometime in the afternoon. She will also notify your producer, who will inform the director. If the show is shooting on the lot, the director may want to see dailies when the crew breaks for lunch or in the evening after they finish shooting.

If the director is able to view dailies at lunch and the producer is able to adjust her schedule accordingly, only a single screening will be necessary. It is most likely, however, that you and your editor will have to attend two runnings. For the director's screening in the evening after shooting, the two of you may have to wait a couple of hours for a director delayed by unexpected problems or, when she is on location, by traffic.

Several years ago some companies began videotaping dailies for directors working on location so that they would not need to come to the studio to view them. The tapes were delivered to their homes where sometime during the evening they could view the dailies on their television sets and then phone the editor if there was something to discuss. It has now become common practice even for some directors working at the studio. Many directors, however, particularly those working on larger TV projects or features, still prefer to view their dailies on a projection room screen where photography and lab work can best be evaluated and where the director and editor can best communicate.

Your film should be delivered to the projection booth at least 10 or 15 minutes before any scheduled running. If you are responsible for transporting the reels, make certain ahead of time that there is a hand cart or other means of transporting them available. Or, if shipping and receiving is responsible for delivery, find out if you have to take the film to the shipping and receiving department or if an apprentice will pick it up from your cutting room instead. Give the people at shipping and receiving as much advance notice as possible about any scheduling changes and about when your dailies will be ready for delivery.

As was mentioned previously, be sure to take with you to the screenings your code sheets or a similar listing of your assembled takes. Other items you should bring are the script notes, camera reports, a lightboard with a pen or pencil and, at some companies, lab reports. A lightboard is a clipboard with an attached battery-operated light. Always pretest it to be sure the light is in working condition and to give yourself time to change batteries or bulb if necessary. Some assistants carry a penlight as a backup. You might also ask your editor if she wants you to take the photocopy of the lined script pages that were shot the previous day.

DAILIES SCREENED

It is still the second day of shooting and your first day of doing dailies. You and your editor are on the way to the assigned projection room to screen your first dailies. En route mention any missing picture or track or temporary print in the dailies that will later be replaced by a reprint from the lab. Also remind her about the outtake, mistakenly printed, that she instructed you to include in the dailies and about the scene on which you received a lab defect report. Clarify whether the editor wants to announce these items at the proper time during the screening or whether you will be responsible for doing so.

Whether you or an apprentice takes the film to the projection booth, you alone are responsible for all the reels being there and being properly lined up for the projectionist. Before you enter the projection room go into the booth, introduce yourself to the projectionist, and if your reels have not been delivered or any are missing call shipping and receiving and try to locate them. Your film *must* be there and ready to start on time.

Check to see if the reels are properly lined up. They should be in consecutive order on a film rack left to right beginning with reel 1—picture and track, reel 2—picture and track, and so forth. If instead of a rack there is a table, the film should be piled top to bottom, beginning with reel 1 at the top.

Even though the projectionist may already know the film is for television or that only television projects are run in his room, tell him you want a *TV aperture,* which is a 1.33 lens. (*Feature aperture* is 1.85 and *wide screen*—Cinemascope, Panovision, and so forth—is 2.35.)

If you have been instructed to work the communications panel in the projection room, arrive early enough to have the projectionist tell you how to operate it. The panel is really quite simple. Some have only an intercom on which to relay directions to the projectionist. Others may also have switches that will activate the film to run forward or backward or to stop. Dailies are usually straight run throughs without any stopping or reversing.

If you do not have to sit at the panel, try to sit next to your editor or as near to her as possible so that you can note the comments made by her and the producer or director. Sometimes a director, producer, or someone else takes seats on either side of the editor unless the editor explains she needs the assistant in one of the seats. Therefore whether you sit next to your editor depends on how strongly your editor saves the necessary seating for you and on how many people are invited to the dailies. Most studio projection rooms used for dailies have limited seating, only about two dozen seats. The director may have invited members of the crew as well as some of the lead performers, while the producer might have invited associate producers, the writers, and a secretary.

If you have been given the responsibility to do so, you will bring to the attention of those attending the screening

1. any missing picture or track,
2. an outtake,
3. a reprint coming, or
4. a lab defect report.

The proper time for this is when the slate of the take involved appears on the screen. In the case of a missing picture, you would make the announcement at the slate that immediately followed the missing scene had you received it. For example, imagine that you did not receive 25D–3 in your dailies and had to order it. At the end of 25C–2, while the slate of 25E–3 is on screen, you would announce that 25D–3 was not printed and that it is a closeup of Bellamy.

The principal notes you will be taking during the screening concern:

1. *Comments,* such as a request that you order an outtake or editorial suggestions to your editor. Examples of the latter would be to try to use a certain reaction or a different angle or when two takes of the same scene have been printed, for example, to use the beginning of one take and the end of the other. Note whether the editor, producer, or director made the comment.

2. *Defects,* other than those reported on the lab defect report such as poor color, out of focus, mike in scene, poor sound, scratches on film, and so forth.

3. *Out of sync.* Hopefully it will not happen to you on your first dailies, but as with all assistants it will happen eventually. It is not usually the result of outright care-

lessness but rather due to an understandable misinterpretation of a technically flawed clapstick or pop. You will probably have an accurate feeling that the only person in the projection room who understands is your editor. And the out of sync always seems to be on a long, long take, thus prolonging your embarrassment. Remember to correct any such errors before coding!

4. *Stock* or production material that can be reused, which is of considerable value, especially on a TV series where certain locations (or establishing shots) and identical actions can be seen repeated episode after episode. Exterior shots of an office building in a lawyer series, a precinct station in a police series, or a residence in a family show are typical examples of location stock. Car runbys left to right and right to left, car chases and crashes, planes taking off, flying and landing are common action stock.

A stock shot can be a combination of establishment and action such as a car runby to camera pulling up to the curb where a major series character steps from the car and walks into a building as the camera pulls back to reveal a full shot of the building. As a matter of fact, this single stock shot has the potential of being used as several individual stock shots in a series.

The above stock examples should also be classified by night or day and by weather conditions such as sunny, cloudy, rainy, snowy, or windy. Any vehicle should be identified by make, model, and color; and individuals, by their costume. If a performer is wearing identifiable apparel that she could not possibly wear in any other episode, that take should not be considered for stock.

Individuals in stock shots must be restricted to regular or *running* series characters. Guest stars and other actors cast in a particular episode may not be used in a stock shot in another episode unless they are given additional compensation and rarely is there any valid reason for such expense.

Sometimes a *portion* of a take has stock value and should be noted. Whenever you have a question about whether a take should be saved for stock, ask your editor at an opportune time. If she can spare the time, the film librarian assigned to your series will attend dailies and will also note the stock. After dailies, you should compare your notes with those of the librarian.

If you are in a situation such as with an independent company with offices and cutting room in an office building where there is not a film library and a librarian is not assigned to your series, you or one of the other assistants on the series will be assigned the responsibility of maintaining the series library, which involves cataloguing and maintaining it as it increases in size. You will also be responsible for requesting and ordering stock from outside libraries.

5. *Music takes.* Those that you will have to identify are the scenes that contain anything musical on screen including dancing, singing, or playing musical instruments. You would not concern yourself with *off-screen music* such as a theatre audience listening to a musical stage number not visible in the shot or *source music* such as a car radio playing music or a scene in a bar including a jukebox providing background music. Again should you have any doubts as to whether or not a take in your dailies should be marked for *music,* always check it out with your editor.

6. *Gag* material, now popularly known to television audiences as *bloopers.* These are humorous (and sometimes obscene) blunders by the actors or technical errors by the crew that are accumulated and eventually edited into a *gag reel,* to be viewed by an eager audience of performers and crew at a Christmas party or *wrap* party at the end of the season's shooting. Occasionally a director will have a take that otherwise would have been an outtake circled and printed only for the gag value to entertain those viewing dailies; you, of course, will add that to your gag material.

CODING DAILIES

Immediately after dailies, if you have an out-of-sync take in any reel, you must correct it before another screening or taping or before that reel is coded. Please remember that after resyncing the incorrect take, you will then have to resync the following take as was explained in the last chapter.

Your major concern will then be to get the film coded so that you can break it down and make it available to your editor for editing. However, when you can get it coded will depend on several factors, among them whether the producer or director, who may not have been able to attend the first screening, is scheduled to see the dailies at the studio later in the day or whether you have instructions to tape the dailies and send the tape to the director's home for viewing that evening.

In each of the above situations you will have to consider as the most expedient option the one that will cause your editor the least delay. If you have time before the next screening and can do so without endangering it, you might decide to get the coding done immediately. It may also depend on the work schedule in the coding room and on the ability of its personnel to do your coding in the required time. Since the apprentice will always automatically leave film tails out after coding, if you are going to screen dailies afterward or if you have to send them out for taping, instruct the apprentice to rewind your film so that they are heads out.

At some time, either between screenings or while your film is being taped or coded, you should complete your code sheets, making them neat and legible so that anyone, particularly your editor, can read and understand them.

You should also have time to make up a *wild track* code sheet for your first episode. Wild track (WT-25) that you removed from the track dailies and placed on your shelf when you were popping tracks must be coded and recorded. Unlike your regular code sheets that begin anew each day, all wild tracks for the entire episode are logged on one code sheet and are under one sequence of code numbers.

You should select a code number that is in no danger of duplicating your regular code numbers such as AH–1000, AJ–1000 or AK–1000 still retaining the beginning, *A* to identify it as the first episode. On the second episode it would begin with a *B* and so on.

In the truncated wild track code sheet illustrated in Figure 10–1 two additional days of wild tracks have been hypothesized to show you that the end code number of a day's wild tracks is spot coded as the beginning number of the next wild track shot the following day.

Columbia Pictures Television

EDITORIAL DAILY FILM CODE RECORD

WILD TRACKS

PROD # 8191 | TITLE: Murder, Anyone? | DATE: _ / _ / _

USE BLACK PEN OR PENCIL ONLY - MAKE ONE CARBON COPY - HOLD COPY FOR SOUND FX

DATE SHOT	SCENE NUMBER	SHOT NUMBER	CODE NUMBER	NEGATIVE KEY AND NOTES	NEG ROLL CAN NO.	SOUND ROLL No.
10/4	WT 25	-1	AK-1000	Sc. 25- Marshall's off-screen dial		14
10/5	WT 10	- 1	76	Sc. 10 - Factory b.g.		17
↓	↓	- 2	128	" " " "		↓
10/7	WT 65	- 1	169	Sc. 65 P.U. Andrews line: "She won't talk"		22
10/8	WT 5	- 1	193	Sc. 5 RYAN: "KIRKWOOD! OVER HERE -- OVER HERE, BY THE TOOL SHED!"		25
↓	↓	- 2	217	Sc. 5 RYAN: POLICE! STOP, OR I'LL SHOOT!"		↓
			243			

LIST ALL WILD TRACKS - DELIVER WILD LINES TO LOOPS AND WILD FX TO SOUND FX, WHEN THE WORK PRINT IS TURNED OVER TO SOUND FX & MUSIC

ASSISTANT EDITOR Erik Rose EXT NO. 1427 PAGE 1 OF 1

-74-7-?

Figure 10–1. Wild track code sheet.

ACMADE CODING

If the company has provided an Acmade machine, you have flexibility in using various tape colors that you do not have with the standard machine. In addition to these colors you can experiment with the available letters to denote the material such as *m* for music takes and *w* for wild tracks until you find a combination that works for you. Get suggestions from other assistants and use the one that appeals to you or try various ones as a starting point from which you can be creative and most efficient.

BREAKING DOWN DAILIES

Breaking down dailies refers to the process of separating each individual take and combining picture and track into a single roll. Give the editor these rolls heads up. You can accomplish this by breaking down the reels from tails to heads using a flange.

Until a few decades ago the identification of each roll was inscribed at the head of the film with a white grease pencil, and the roll was secured with a rubber band or paper clip. Additional information for the editor such as comments from the projection room was written on notepaper. In the late 1950s or early 1960s some unrecognized genius proposed that a card be used to provide a more convenient and legible area for all the required notes. This card is called a *tab* or *trim tab;* it is 4 × 1¼ inches with a punched hole at the top so that it can be hung from a hook. It is attached to the roll by a paper clip or rubber band.

This single tab begat the double tab, which is the same length but double the width so that it can be doubled over (see Figure 10–2). I prefer the double tab for the obvious advantage of providing more space for information. There is another advantage too as shown in Figure 10–2; the double tab can also be used to secure the film roll without requiring a paper clip or rubber band.

There are three primary items that must be included on the tab:

1. scene/take number that is also inscribed on the film at the clapsticks and pop;
2. description; and
3. beginning and end code numbers.

Other items to be included are the production number and remarks from your code sheet such as MOS, music, gag, stock, and so forth. Using abbreviations, you will be

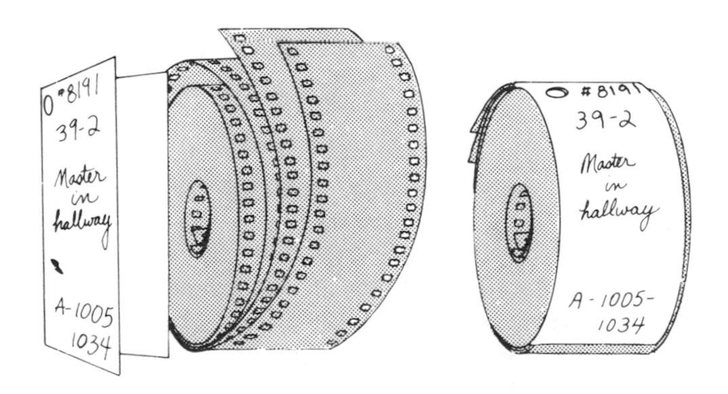

Figure 10–2. Tabs.

able to include most of them on the front of the tab. You have the use of other sides of the tab should you need it for lengthy comment.

Descriptions can usually be brief, such as *master, 2-shot, o/s Mary* or *C.U. John.* Some editors may request more detailed descriptions, while others will simply ask you to leave sufficient space for them to write their own descriptions. I prefer the latter and ask my assistants to use a fine-tip black pen to inscribe the scene/take number at the top of the tab and the code numbers at the bottom, leaving the space in the middle for my description, which I write in red as I run through the rolls of a scene. (*Scene* is loosely used either to refer to a single take, as in scene 25, take 4, or to mean all the takes—25-4 through 25N-1—of a particular scene in the script.) I may also include on a tab other editing notes about things an assistant could not possibly have anticipated.

When you consult with your editor about how to prepare the double tabs, ask if she prefers the punched hole on the upper left or upper right because that will determine the side of the tabs on which to write the primary information. Have the tabs made up before your film is available to be broken down.

When you receive the dailies following coding, you can choose one of two possible methods of recording reel code numbers. You can rewind the picture reels to their heads, and winding down head to tail without the synchronizer, you can read off the code numbers now on the film.

A variation is using the synchronizer and the footage counter as a double check. The advantage of this method is that you are concentrating on only one task, recording the code numbers on code sheets and tabs. With experience at using the sync machine you can learn to do several reels simultaneously.

A second method is to note your code numbers as you are breaking down the dailies. This method is obviously more difficult and becomes feasible only after you become skillful at breaking down the dailies. Assume therefore that, for a short time at least, you will be using the first method.

To break down the dailies using either method place a reel of picture and the accompanying track on the left rewind. Wind broken-down dailies onto the flange and remove and file them as rolls of film. Wind the picture and corresponding track on top of one another with picture on top of the track unless your editor prefers the picture under the track.

Break down the dailies from the tails, first checking the end syncs to be sure coding is correct. Use a female flange or a male flange with adaptor on the right (take-up) rewind. Place the end picture code over the track code as you do when checking the end syncs, and cut both with your scissors at a point a couple of frames past the last frame of picture. Discard the end trims and engage both picture and track together on the take-up flange with track under picture. When you get to about a foot beyond the clapsticks, match the code numbers as you did before and cut the film. Still matching code numbers, go on to a frame or two after the end frame of the next take and cut the film thus discarding the trim.

Release the roll from the flange, secure it with its double tab, and place the roll on your bench shelf. Line up the daily rolls left to right on the shelves, on wooden roll racks if you have them, or in open trim boxes on reel racks in numerical and alphabetical order so that you can easily and quickly find a particular scene that your editors wishes to cut.

Save the head leaders for future dailies. Remove the reel number and code number tapes so that you will not neglect relabeling them correctly later.

RED-LINING STOCK

Any takes you may have marked for stock during dailies should be *red lined.* This means that you have to inscribe a red line on the right, or clear, edge of the picture for the entire length of the take. You can best accomplish this by placing the picture in your synchronizer as you are breaking down the take and, at the same time as you run it through the machine, holding a large, wide-tipped red marked against the film edge. This will permanently identify the take, or any portion of the take, as stock. You should also write *stock* on the tab. Stock will be discussed in greater detail in Chapter 14.

MUSIC TAKES AND MUSIC CODING

Those takes you identified in dailies as music takes are broken down using separate rolls for picture and for track. The individual (music) picture rolls can be placed in trim boxes marked for the music editor. The individual (music) tracks that *include dialogue* can be placed in numerical scene sequence on the shelf with the other dailies. The other tracks, containing only music, will never be used by your editor and can be filed in a trim box marked *Production music tracks replaced* with scene numbers identified.

In order to cut this single music scene the editor needs to have the music picture takes synced and music-coded to a copy of the master music playback for that scene. Send the picture takes to the music editor assigned to your series. He will have a *master playback* of the music already coded with a number like KJ–0000 at the beginning of the music. He will lip sync the various picture takes with the playback and will mark each one with an appropriate beginning number for spot coding.

For example assume your scene includes a master, medium (med) shot and close-up (C.U.), all having the beginning of the music. All of these angles may have some action or dialogue before the music begins. Nevertheless, the music editor marks the start of the music to be spot coded KJ–0000 on each of the three scenes. It is a three-minute musical selection and the master and closeup run full length, so that their end code numbers are KJ–0270 (270 feet = three minutes of film). However, the medium shot was halted partway through and goes only to KJ–0147. Beginning about 12 feet ahead of where the medium shot ended, a pick up (P.U.) was shot, and the music editor, by lip syncing it will mark it to be spot coded KJ–0135.

In this manner she will mark all the takes you have sent and will then send them, along with a *music playback copy,* to the coding room for coding. You or the music editor will preselect a color (usually yellow) so that the music codes can be easily differentiated from your black daily codes. (More will be said about the role of the music editor in Chapter 16.)

When the spot coding has been completed, you can attach the tabs to the picture rolls and incorporate them and the playback copy with the dailies on the shelf. Those production tracks with dialogue you previously placed there should be rolled up with the corresponding picture.

You should indicate on the code sheets the music codes assigned to the music takes, or if you have a number of music sequences, it might be advisable to make up individual music code sheets. Identify the selected music takes and the beginning and end music codes for each take.

Not until all this has been accomplished can the editor begin cutting the music scene or sequence. It is therefore imperative that you get the film to the music editor as soon as possible, so that she can work on it at the earliest opportunity. Keep advised about the music editor's progress so that your editor will not be held up.

If you are on a project for which a music editor has not yet been assigned, you will have to arrange with the mixer to send the playback with the sound dailies for transfer to 35mm. Either you or your editor will then have to sync 35mm playback and picture and mark them for music coding.

MOS TAKES

Be sure to indicate MOS takes on your tabs. You should confer with your editor about whether she wants you to roll up track leader with these silent takes or to separate them and give her only picture. In the latter case, save the leader for future use.

The majority of MOS takes involve a sequence of many *cuts* such as those found in a chase or a fight scene. Most editors prefer to work with picture only, editing the entire silent sequence or section of a scene without worrying about matching leader. When the editor has satisfactorily completed the work on this sequence, she can line up one piece of leader in its entirety.

PREPARATION FOR EDITING

The fascinating reminiscences of the eminent New York film editor Ralph Rosenblum were published in 1979 under the curious title, *When the Shooting Stops . . . the Cutting Begins.* In the Introduction (p. 5) he states that "A feature-length film generates anywhere from twenty to forty hours of raw footage. *When the shooting stops,* that unrefined film becomes the movie's raw material, just as the script had been the raw material before. It must now be (edited) . . ."

This is a perfectly sound statement but it was the only reference I could find to the title of the book—and that title is a misnomer. I have to assume that what was implied was that when the shooting stops, the *recutting* begins. In other words, that is when the vital editing by the editor in collaboration with the director and producer begins. Whether it be a feature, long-form television, or a series episode, however, the editor actually *begins* editing as soon as possible *after the first day of shooting* with limited input from these collaborators until after shooting is completed. If *When the Shooting Stops . . . the Cutting Begins* were true, most projects would not make their air dates or booking schedules.

The impending weekly air dates make it imperative to begin editing episodic series as soon as possible. Your editor must be ready, for example, for a first cut screening no more than six days following the last day of shooting—and sometimes certain circumstances cause this date to be moved up to as early as one or two days following the end of shooting!

If you are able to break down any portion of your dailies containing a *complete* scene or sequence, the editor may begin cutting the afternoon of the second day of shooting. If, however, you are unable to break down any dailies until late that day, she will most certainly begin editing early on the third day and most assuredly *not* "when the shooting stops." Your editor will want to start on a complete scene in which all *coverage* has been shot. *Coverage* consists of the various angles such as the full shot, two-shot, and closeup shot on a particular scene.

Looking at the code sheets (Figures 9–3 and 9–4), what do you think your editor will ask for? She will certainly not want scenes 30 or 46 because they are obviously isolated scenes. Your editor will probably prefer to cut scene 39 in the brief time remaining that second afternoon, and the next morning will begin editing the longer, more difficult scenes 24 through 25N. Do not give your editor more or less than she requests. She may tell you which scenes she would like to begin editing before you start to break down dailies. You should try to get the reels containing those scenes coded first and to break down those scenes as soon as possible.

Before you tell your editor that the dailies are ready, make certain everything is there. Are there any individual takes such as music, a missing track, a picture reprint, or a wild track still in the coding room? Have you incorporated all such items in their proper places? Do not forget about any long takes which you had to put on reels. Exactly where does your editor want those reels placed, and on which of the editor's shelves do the rolls go?

Since your dailies should be lined up in numerical and alphabetical order, it will be a simple matter to move them in that same order onto the editor's bench. By placing pressure on both ends of a stack of rolls, you can learn to move a considerable number of them at one time. But, if a middle roll should suddenly drop out, you will find yourself in an embarrassing game called *pickup*.

Should the center of a single roll drop out, thus unrolling the film, it will not be embarrassing—it will be downright frustrating and very time consuming to reroll. Put the roll in a trim box and, if you are lucky, perhaps you will be able to wind the ends of the take back into the center. Usually, you will have to use an alternative method such as cautiously standing on a chair or bench, unrolling the film into a bin, and lifting the film as high as possible and carefully shaking out all the twists, kinks, and tangles that can severely damage the film. When you have smoothed out the film from end to end, you will be able to reroll it. If you are a tall person, you may not need the furniture and will be able to shake out the film and reroll it at the same time.

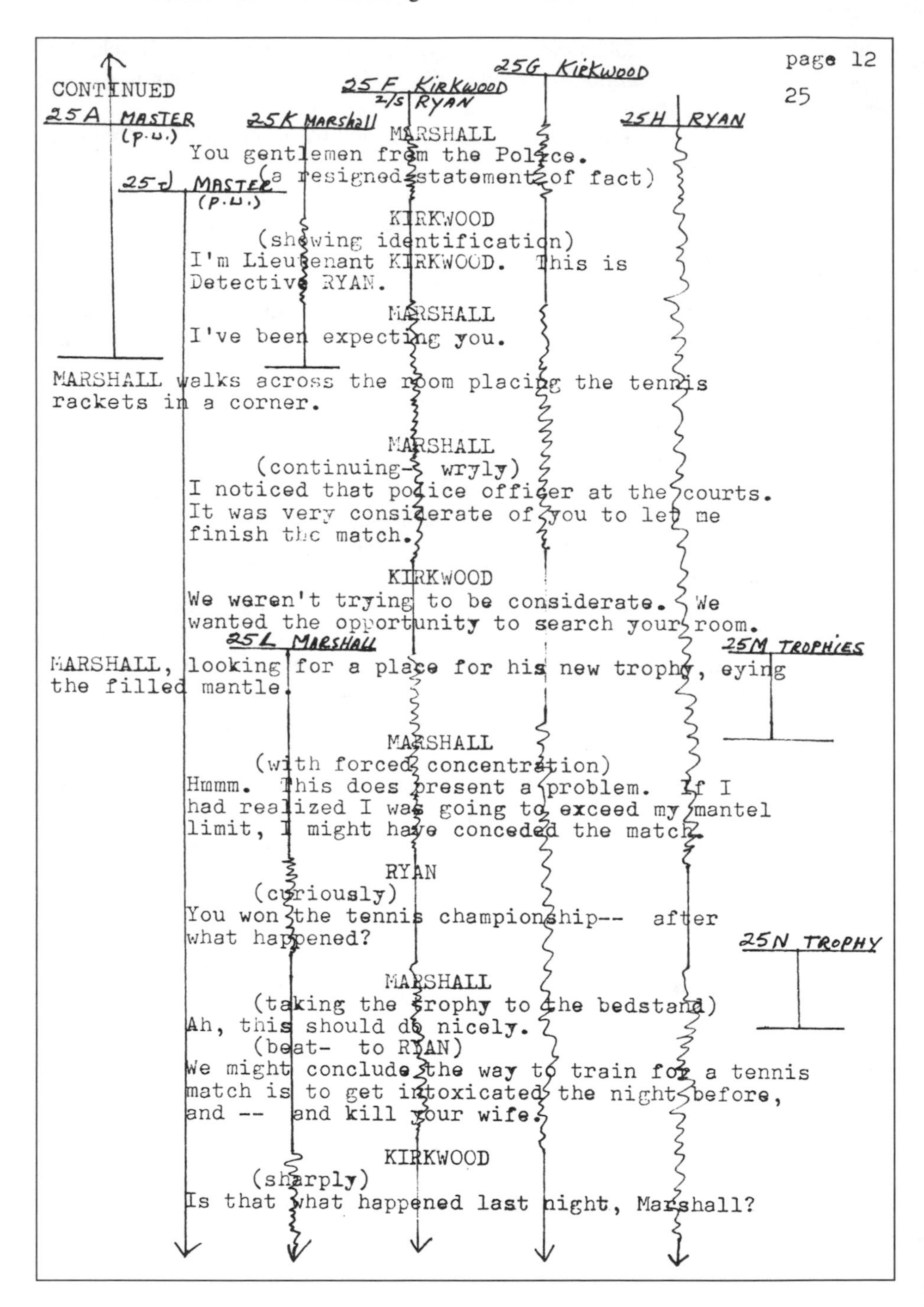

page 12

25

CONTINUED

25A MASTER (p.u.)
25J MASTER (p.u.)
25K Marshall
25F KIRKWOOD 2/S RYAN
25G KIRKWOOD
25H RYAN

MARSHALL
You gentlemen from the Police.
(a resigned statement of fact)

KIRKWOOD
(showing identification)
I'm Lieutenant KIRKWOOD. This is
Detective RYAN.

MARSHALL
I've been expecting you.

MARSHALL walks across the room placing the tennis
rackets in a corner.

MARSHALL
(continuing-- wryly)
I noticed that police officer at the courts.
It was very considerate of you to let me
finish the match.

KIRKWOOD
We weren't trying to be considerate. We
wanted the opportunity to search your room.

25L MARSHALL
25M TROPHIES

MARSHALL, looking for a place for his new trophy, eying
the filled mantle.

MARSHALL
(with forced concentration)
Hmmm. This does present a problem. If I
had realized I was going to exceed my mantel
limit, I might have conceded the match.

RYAN
(curiously)
You won the tennis championship-- after
what happened?

25N TROPHY

MARSHALL
(taking the trophy to the bedstand)
Ah, this should do nicely.
(beat- to RYAN)
We might conclude the way to train for a tennis
match is to get intoxicated the night before,
and -- and kill your wife.

KIRKWOOD
(sharply)
Is that what happened last night, Marshall?

Figure 10–3. Lined script.

LINING UP BY LINED SCRIPT

If you suggest to your editor that you line up the scene she has asked for in *lined script order,* your editor will undoubtedly be surprised and pleased that you, a novice assistant, might be able to do this whether or not she actually lets you do it.

To do it you must understand a lined script and how most editors use it in their editing. The lined page illustrated as Figure 10–3 is a continuation page of scene 25. The script supervisor *lines* each printed take indicating where it begins and ends. Obviously this is a single camera scene. Otherwise the lined script as well as all your daily forms would have indicated *A* camera, *B* camera, *C* camera, and so forth.

The straight line designates *on-camera* while the wiggly line means *off-camera* dialogue. The two master (pick-up) takes, 25A and 25J, have only straight lines because they are full shots with everyone on camera. The two-shot of Kirkwood and Ryan has straight lines when either of them is speaking and wiggly lines of Marshall's dialogue because in this two-shot he is off camera. The two takes of Marshall's closeup, 25K and 25L, have straight lines on his dialogue and wiggly lines on the other off-screen characters.

The common editing approach is to consider all coverage of dialogue and action in script continuity. Should you give your editor this scene in alphabetical order, she will rearrange it in lined script order: 25A, 25K, 25J, 25F, 25G, 25H, 25L, 25M, 25N.

The rationale for this sequence of takes is that the first dialogue on this page is Marshall's. There are only two takes for this speech. The master, 25A, and Marshall's closeup, 25K, would be lined up first. The next dialogue is Kirkwood's. Scene 25A roll is already on the shelf but another pick up of the master, 25J, begins here, so that would be next. Now the editor selects two other takes that cover this dialogue: a two-shot, 25F, and a *single,* or closeup, of Kirkwood, 25G.

What would you choose next? Although Ryan's closeup begins at the top of the page, he does not have any dialogue until near the end. Marshall's pick up, 25L, and an insert, 25M, precede it. In this case, however, instead of following sequential dialogue your editor chooses 25H and then Marshall's pick up and the two inserts. Why? It is simply a personal choice based on the possibility that somewhere during that ¾-page before Ryan speaks the editor might want to use a piece of the take to show Ryan reacting to what is being said. It is assumed that the performer (Ryan), being shot while others are speaking off camera, is, or should be, reacting to their dialogue.

Note that when you have multiple choices for a particular line of dialogue at the beginning of a scene such as a master, a medium or group shot, a two-shot and possibly a medium closeup, or a head closeup, most editors want their rolls lined up from wide angle to close shot.

Do not be disturbed if, after you have lined up a sequence in lined script order, your editor makes a few arbitrary adjustments. You cannot anticipate some of these adjustments, but the more you work with your editor the more you will be able to predict her preferences and, more important, the more you will begin to understand the use of the lined script in editing.

Finally, the lining up of the rolls on the editor's shelf does not necessarily indicate the order in which the takes of the scene will be cut. This is only the procedure of putting the *choices* in an organized sequence.

Unfortunately not all lined scripts are as neat and explicit as the one illustrated. A few script supervisors do very sloppy work. They do not identify the coverage of the takes; they will just line each take down the page without any reference to on-screen or off-screen dialogue or action so the pages are of little use to the editor. Or their notes may simply be illegible. It may then be incumbent upon the assistant first to go through the script pages providing the identifications and correcting the lines before the editor begins cutting.

A sloppy supervisor is usually unreliable in other ways. She may not take a few minutes at the end of the day to make certain that camera and sound reports agree with the circled takes on her reports. Entreaties from editorial may receive a frivolous response. Pleas to producers to intervene are often met with unbelievable indifference, simply because they do not fully comprehend the unnecessary extra work and delay supervisory sloppiness may cause in editing. Consequently, having done all you could, you and your editor must just bite the bullet and suffer.

Fortunately, most script supervisors are very conscientious and are immeasurably helpful and cooperative. They usually do not receive the recognition they deserve.

SCRIPT LOG

On the opposite page, the reverse side of the preceding script page, the script supervisor lists the takes of each angle. I call this list the *script log* (Figure 10–4). It is important that you understand the difference between it and what we call the *script notes* (see Figure 8–1). The log, like the camera and sound reports, lists *all* the takes shot whereas the script notes, which are so vital to the assistant in completing dailies, list only the circled takes. Unfortunately there is no conformity in the industry in the use of terms for these two forms. Some companies transpose the terms.

As you can see, the script log offers a more detailed description of the takes than do the one-liners in the script notes. It also gives the timing on each shot and by circling a take specifies that it is a *print.* Other noncircled takes can be described as *comp* (complete), *inc* (incomplete) and *hold,* the latter meaning that, although it is not being printed, it is nearly as good as the circled take.

The script log is particularly helpful to the editor when she is considering ordering an outtake. If, for example, she felt displeased with a piece of dialogue or action beginning at 50 feet into scene 25J–4 and was obliged to use that angle, by studying the log she would know that her only alternative was to order 25J–2. Why not order take 1 or take 3? Take 1 is timed at :25 or 37½ feet and take 3 is timed at :10 or 15 feet. Obviously, neither take covers the action at 50 feet.

Some editors will require that you note code numbers for each take on the script log or on the lined script. Usually this is required on the longer projects.

10/4 25F		
1- 1:12 inc 2- 1:07 inc 3- :27 inc (4)- 2:50 PRINT	Wide 2/s as KIRKWOOD moves to door right, Ryan at door left – MARSHALL ENTERS AND TURNS back to cam AS KIRKWOOD CLOSES DOOR – dial as KIRKWOOD + RYAN MOVE to FACE MARSHALL FOR CLOSER 2/S	
10/4 25G		
(1)- 2:55 PRINT	Med/s KIRKWOOD at door to C.U. as he moves to cam (said between 1 and 2")	
10/4 25H		
1- :15 inc (2)- 2:52 PRINT	Med/s RYAN at door, moves to cam for C.U.	
10/4 25I	P.U. C.U. KIRKWOOD	
(1)- :15 PRINT		
10/4 25J		
1- :25 inc 2- 2:48 Hold 3- :10 inc (4)- 2:46 PRINT	MASTER, diff angle, KIRKWOOD/RYAN ENTER RIGHT to MARSHALL WHO MOVES to REAR, places RACKET against WALL, then MOVES to MANTEL and bedstand (TROPHY in left hand, RACKET under RIGHT ARM – TROPHY to RIGHT Hand)	
10/4 25K		
(1)- :05 PRINT	C.U. MARSHALL, EXITS RIGHT REAR	
10/4 25L		
1- :56 inc, (2)- 2:20 HOLD (3)- 2:22 PRINT	MED/S MARSHALL ENTERS RIGHT to MANTEL – pan to bedstand, back to KIRKWOOD/RYAN	
10/4 25M		
(1)- :10 PRINT (2)- :08 PRINT	MOS 1-Pan C.S. trophies L-R, R-L 2- Widen C.S. trophies	
10/4 25N		
(1)- :05 PRINT	MOS C.S. Bedstand, MARSHALL's hand enters R, setting down TROPHY	

Figure 10–4. Script log.

ORGANIZATION OF RECORDS

On the third day of shooting you will receive the second day of dailies, and the procedures described in previous chapters will be repeated from the syncing of dailies through breaking them down. Before you receive the dailies and their related papers, you should have the first day's records filed away.

There is no one standard method of filing. The materials you receive are several three-ring notebooks and several clipboards that can be hung on a wall near your desk. You will usually have to provide your own single punch, three-hole punch, and stapler.

The filing system I recommend is using one notebook as a *code book* containing only code sheets and script notes, each in a separate section and in shooting day order. Another notebook could be used for the other forms—sound, camera, sound transfer, and lab reports and delivery slips. These forms could also be kept individually in shooting day sequence. Some to these forms, such as the camera and lab reports and delivery slips, may be too small to be conveniently placed in a three-ring notebook, but by securing them with staples, fasteners, or rubber bands you can put them in small pockets in the pocket at the rear of the notebook. If you prefer, these smaller forms can be placed on two-ring clipboards or into two-ring notebooks.

Some assistants like to file all the forms together for each day of shooting. I do not care for this method because if, for example, you are looking for a particular sound report you have to flip through all the other forms to find it. In the final analysis it is really up to each of you to select your own filing system. Whatever works best for you is the one you should use, keeping in mind, however, that it must also work for your editor.

EXCEPTIONS (Not generally applicable to episodic TV)

Kem Dailies

As was previously mentioned, at present I know of no prime-time episodic series edited on a flatbed. It is really impractical as compared with the moviola considering the cost, the space required, and the personnel needed. But you should be aware of its operation should you be assigned to a miniseries, TV movie, special, or feature where a flatbed is often used.

Although Kem dailies are usually assembled as any other dailies, the process is quite different beginning with the viewing of the dailies. As was mentioned in the previous chapter, daily screening can be done on the flatbed itself in the cutting room at the producer or director's discretion, rather than maintaining the expense and restrictions of a scheduled screening in a projection room. Viewing dailies on a flatbed affords the producer and director considerable latitude and eliminates a lot of film schlepping.

After the viewing, instead of getting the coding done you will first have to reassemble the dailies according to the requirements of the flatbed (see Figure 9–5). Remember that the flatbed uses Kem rolls on cores usually with a number of takes on each roll. These rolls sometimes do not provide the ready availability of individual daily takes broken down for the moviola so your objective is to try to anticipate the editing of a scene in a general way so that the editor will do a minimum of running back and forth in one roll to make a series of cuts. It will be easier for the editor to make those cuts if she is able to go from a certain point in one roll to a point in another roll, and then back and forth between the rolls.

Donn Cambern, A.C.E., eminent Hollywood feature editor of such films as *Easy Rider, Romancing the Stone* and, more recently, *G-men,* edits on an eight-plate Kem Universal with two sound heads and two screens. With four plates on the left and four plates for take-up cores on the right, Cambern can have two rolls of picture and two rolls of corresponding track on the machine at one time. With a flick of a switch he can go from one roll to the other. With two screens he can have his selected cut on one screen and look for his *matching,* or successive, cut on the other screen.

His dailies are viewed in *shooting sequence.* Then assistant editor Saul Saladow reassembles the film in a *checkerboard* pattern so that regardless of which two rolls are on the machine there will be minimal need to go to a third roll until the editor is also ready for the *matching* fourth roll.

Let us imagine that Cambern's dailies include scene 23 with several different angles for coverage. Also, since this is a feature several takes may be circled on each angle. For example, three takes on the master, 23–3, 23–5 and 23–6, may have been printed. These master takes would be assembled on Kem roll 1, and if there was space on the roll without making it too bulky, they would be followed by the wide over-shoulder shots. Roll 2 would begin with the two-shots followed by *matching* wide over-shoulders. Roll 3 would begin with over-shoulder singles followed by single shots (closeups). Roll 4 would have the matching over-shoulder singles and then the matching closeups.

Cambern uses the Acmade code system. The first three digits identify the scene, 23. Therefore the beginning code number on all the rolls is 023. No attempt is made to try to identify each individual angle such as 23A, 23B, 23C. The middle digit of the Acmade code is left blank and the last four digits represents the Kem roll, i.e., roll 1 is coded 023 1000; roll 2, 023 2000; and so forth. Most scenes do not require more than five or six Kem rolls but should there be a sequence with an exceptional amount of film, then roll 9 would be coded 023 9000; roll 10, 023 0000; roll 11, 023 *A* 1000; roll 12, 023 *A* 2000; and so forth. The use of the *A* for the middle digit denotes the continuation of the numbered rolls and has no relation to scene 23A.

There are various ways of organizing Kem rolls depending upon the equipment being used and the editor's preference. Before you begin reassembling the takes know exactly what your editor would like and proceed accordingly.

CHAPTER 11

THE EDITING BEGINS

It is now the morning of your third day as an assistant editor. On the preceding day you completed your first day of dailies in the following progression. Tracks were popped; picture and sound, synced; dailies, screened then coded and finally broken down into individual take rolls, picture and track together. Rolls were lined up on your bench. Your editor can now select a scene and begin editing. If you have provided the editor with all the necessary equipment and supplies, including leader and missing banners (refer to Chapter 6), you and your editor can now work independently at least for the time being. While he begins the editing process, you can concentrate on your second day of dailies.

Imagine that your editor decides to cut scenes 24 and 25 through 25N (refer to earlier Figures 9–3 and 9–4). Your editor or you will first line up all the rolls for this sequence on his bench using the lined script (Figure 10–3) described in Chapter 10.

THE EDITING PROCESS

The following is a very simplified description of the editing process designed to demonstrate, not any sort of editing technique, but the evolution of *cuts* and *trims*. *Cut,* as a verb, means to *edit.* As a noun, it is used variously to refer to a portion of a take used by the editor, or as a screening of the edited film, as in *director's cut, first cut,* or *final cut.* As an adjective, it means *edited* as in the *cut reel,* or the *cut picture*.

A cut, as used here, is the portion or piece of each take that the editor selects to use in the movie. The editor may use one or several cuts from a single take and these cuts, each interspersed with those from various other takes, are attached one by one to reels on the rewind. Simply put, this is *building* a scene, and the assembling of the scenes is the building of the movie. Those sections of the takes he does not want are the *trims,* and they will be placed on hooks in an orderly fashion in the *trim bin* and eventually filed.

The editor begins by placing two empty reels with a clamp on the right bench rewind. He takes two short pieces of temporary leader, marks them as *scenes 24–25* in black and emulsion-up on picture and in red and cell-up on track, and attaches them to the reels. In addition to identifying the sequence this leader will also protect the film.

The first scene, 24–1, is a full, or master, shot of three men walking down a hallway to an apartment door that the character Bellamy unlocks; then they enter. There are no other angles on this scene. The editor runs the roll of picture and track in the moviola, marking with a white grease pencil where the scene should begin. At the end of the scene, the end of the cut is similarly marked, probably where the men start to enter the apartment.

Picture and track now being tails out, the editor rewinds both together to heads on the moviola rewind. The editor now has all the film for the take together in one roll as it was before running it through the moviola. He now turns back to the editing bench and places picture in the first gang of the synchronizer and the track in the second gang, using the code numbers for sync. Now, using the hand knob on the synchronizer, the editor rolls down to the beginning mark on the picture, marks the exact frame line at that point, and makes a corresponding mark on the track. Both are then cut in the butt splicer on these exact marks.

The picture and track pieces on the right side of the splicer are those that are hung in the trim bin. These consist of the slate and possibly a few feet of the beginning action. This is the *head* trim that the editor places heads up on the extreme left hook of his trim bin (Figure 11–1). The tab for scene 24–1 should be placed on the hook above.

Paper clips are used to attach the cut of the three men walking down the hallway to the leaders, and with the film still in the sync machine the editor winds the scene onto the reels and stops at the end mark, where the men start to enter the apartment. Before cutting here, however, he may decide first to check the exact point at which to begin scene 25–4, showing the men continuing their entrance from a perspective within the apartment (see Figure 8–1, the script notes). Checking this way provides the opportunity to adjust the 24–1 end mark a few frames if it is desired so that a *match cut* can be made. This means that the timing of the two pieces of film are coordinated so that the movement of the men and the transition of action from exterior to interior will be both smooth and realistic. Having finalized this decision, the editor cuts the film at the end mark, and now the first cut is complete and attached to leaders on the reels. The remainder of 24–1 is the *tail,* or *end, trim* and is added heads up to the first bin hook because there will be no further use of this take.

Before continuing, even though this material was viewed in the daily screening, the editor reviews all the takes of scene 25 in the moviola, marking the beginning and end of each selected cut. After this has been done, he returns to the master, or full, shot of the men entering the apartment, 25–4, and cuts it where it was previously marked for the match cut. It is paper-clipped to the tail end of the preceding hallway shot, 24–1, on the reels and the head trim and tab of 25–4 are placed on the second set of hooks in the trim bin.

Now the editor rolls down to the end mark on the master, 25–4, and cuts there. There are now two cuts on the reel. The remainder of roll 25–4 is placed somewhere conveniently on the bench; the editor may wish to use this master again later.

In our example scene what follows is a verbal exchange between the characters Bellamy and Kirkwood. The second cut, of the men entering the apartment, may end with Kirkwood asking Bellamy a question. The next cut is a closeup of Bellamy responding to this question.

Thus the editor turns from the bench, and with roll in hand places the head of Bellamy's closeup, 25D–3, and the corresponding tab on the third hook. The film is unrolled down into the bin until he arrives at the mark on the picture for the beginning of Bellamy's response where the film is then cut. Because the editor first hung the head of this take on the bin hook and rolled the film directly into the bin, the head trim is already in the bin. When there is sufficient film before the cut, this procedure is rather obviously more efficient than first rolling down, making the cut, and *then* hanging the film on the hook.

The editor clips the picture and track heads of the new cut to the picture and track ends of the second cut. He winds down the film in the synchronizer to the end mark on Bellamy's dialogue and cuts there. There are now three cuts on the reels. The remainder roll of 25D–3 is placed on the bench with 25–4 for possible future use.

Kirkwood's closeup, 25B–3, is next since he has the next line of dialogue. The same procedure is followed with the head trim and tab placed on the fourth hook and the cut of Kirkwood added to the reels. (There are now four cuts.) The remainder of this roll is also left on the bench for further use.

The next dialogue is again Bellamy's so the editor retrieves 25D–3, rolls down to the beginning mark for that dialogue, and cuts the film. There is now a *middle* trim for this take, the film between the first closeup cut of Bellamy and this second cut, and that is added to the third hook. The procedure continues essentially in this manner until the scene or scenes have been built.

It must be emphasized that the editing procedure outlined above does not include other editing choices such as the overlapping of dialogue onto reactions of either Kirkwood or Bellamy or the third man, Ryan. The purpose of this discussion, however, is not to try to explain *editing* but instead to enable you, the assistant, to visualize how cuts and trims evolve. Once a single cut is taken from a roll, or take, there will always be a head trim and a tail trim.

Trim does not necessarily imply a small piece of film. It can be any length from one frame to hundreds of feet. Any trim, picture or track, that is less than one foot in length, is termed a *short trim* and should be placed in a short trim box. If a trim is quite large and would unnecessarily overload the bin during editing, it is retained as a *roll* and is temporarily set aside on the editor's bench. If the editor decides not to use anything from a take leaving it intact, it becomes an *out* (not to be confused with negative outtakes, which you might recall are uncircled takes) and it is also set aside on the bench.

ASSISTING YOUR EDITOR

In this initial phase of editing your services have not been required except to answer phone calls. This will allow you to complete the dailies without much interruption. Your help will be needed, however, when the editor finishes cutting the sequence. You will then attach a few feet of temporary leader at the end of both picture and track. You should label these end leaders, for example, *End- Scs. 24–25.* You can then rewind picture and track individually, removing the paper clips and splicing as you do so. Since you are going from tail to head, when you finish splicing, both reels will be heads up. Some editors as they edit prefer to wind up their picture and sound cuts on *one* reel, rather than on two separate reels. In the latter case you have to splice and rewind the picture and track concurrently.

There are some editors who do their own splicing as they cut so that they might periodically check their editing in the moviola as they proceed. Probably they will also rewind the film so that your help will still not be required.

An editor likes to recheck a scene that has been cut as soon as possible because, first, it is still fresh in his mind, and secondly the trims are still in the bin should it be necessary to make any changes. Never wind up the trims until you have the editor's approval.

If the splicing you have to do is extensive or is delayed because you have to prepare the dailies, the editor can begin cutting another scene. If the cut scene has to be shown to a director or producer and the splicing conflicts with dailies, help from an apprentice or another assistant may have to be obtained even though you, like most assistants, prefer to do the job yourself.

THE TRIM BIN

35mm trim bins vary in size (see Figure 11–1). The largest bin, which is approximately 35 × 24½ × 27 inches with 17 sets of hooks, is preferred, of course, because it will

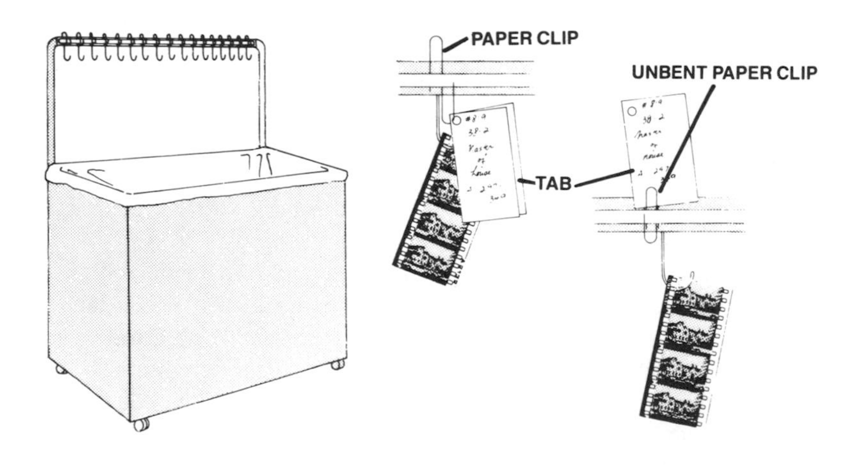

Figure 11–1. Trim bin.

accommodate the most film and the most takes. The largest bins will also take up more of your limited space, but they are on wheels and are easy to move about. Whenever the editor has a scene with more than seventeen takes then, once all the seventeen hooks in the first bin have been used, it will be necessary to continue into a second bin.

In the foregoing it has been assumed you have bins with double hook racks so that you can use the lower rack for the trims and the corresponding tabs can be hung directly above from the upper rack. You may have bins lacking this upper hook rack, however, leaving the editor with only one row of hooks for both film and tab. If the editor places the tab on a hook before he places the first (head) trim on it, that trim and succeeding trims will cover the tab. He needs to see that tab at a glance to guide him in placing the right trim on the correct hook.

Therefore, if your bins lack upper hook racks, you have to devise upper racks for the tabs. This can be accomplished by using strong one-inch cloth (camera) tape to secure two-inch paper clips along the length of the rack above the hooks. The clips should be attached vertically so that the tabs can be clipped into them, or, alternately, a prong on each clip can be bent outward so that the tabs, which have a hole punched at the top, can be hung on the prongs. You should prepare all the bins in this manner before your editor begins cutting.

WINDING UP TRIMS

After the editor has rechecked the cut sequence and made any changes, he will indicate that the trims may be wound up. You should do this as soon as possible, particularly if there are some trims in the second bin, because an empty bin will be needed for the next scene that is cut.

First, get any *outs* or *rolls* from the sequence that the editor has set aside on his bench. Since outs are complete, uncut scenes ready to be filed, set them temporarily aside on *your* bench. The *rolls,* which should already be represented by a tab and related (shorter) trims in the bin, will have to be wound down into the bin, and each one should be placed on the proper hook.

As you already know the trims are hanging from the hooks heads up. Your objective is to wind up each take, picture and sound, into one trim roll tails up. Sounds easy, doesn't it? But there is more to it than just that. Try to envision a completely full trim bin such as the scene 24–25 sequence. Since the editor began hanging the trims left to right, your inclination might be to begin winding up with the first trim on the left, scene 24, and continue left to right. Since the ends of the earliest trims will be at the bottom of the bin underneath all the other trims, you will be pulling up against the weight of the other trims and increasing the danger of tearing either the trim you are winding up or another trim. Although that danger is always present, you will minimize it if you begin with the trim on the extreme right hook and continue right to left. Also as you are winding up one trim and carefully pulling it through the other trims, there is the other danger of yanking some of the others off of their hooks. This can be prevented by securing them with rubber bands stretched flat across the hooks and releasing one hook at a time as you wind up.

Before winding up a trim, you must flip through each piece and make certain it has been placed on the correct hook. Some editors are very careless and even with the help of the tabs constantly file their trims incorrectly. And even careful editors sometimes err. By checking description and code numbers as indicated on the tabs you can be precise and accurate. If you carelessly permit a trim to be rolled up into the wrong trim roll, it may cause you and your editor many minutes of frustrated searching later on and may even jeopardize a screening.

Some editors prefer that a trim be wound up with track trims beneath each corresponding picture trim. Other editors insist that the track trim be on top of the picture. Some editors also want the lowest code number of a trim (the head) placed first graduating to the highest code number (the tail). This is the reverse order to that the editor used in placing them on the hooks. Make certain you are advised of any preferences your editor may have before you begin winding up the trims.

Once you are certain that all the pieces of picture and track of a trim are correct and that tracks and code numbers are in the position and order your editor wishes, you may wind up that take into a trim roll. Grasp the trims firmly so that as you release them carefully from a hook no trim slips out and drops into the bin. Bind the trim together with rubber bands (the number of rubber bands will depend on the amount of trims) or a paper clip according to your editor's preference, and wind it up on a flange heads to tails as a *trim roll.* It is now tails out. The roll can now be secured with its tab. When you have rolled up all the trims, always check the bottom of the bin to see if any trim has accidentally fallen there. Any trims you find loose in the bin have to be replaced in their proper rolls. Since the rolls are tails out when you unwind a roll down into the bin, the bound center of the roll will end heads up, ready for you to insert the loose trim in the proper place.

FILING TRIMS AND OTHER FILM

Trims and other film not being edited into the picture are eventually filed into cardboard boxes such as those containing your picture dailies. As previously suggested, you will also get a supply of the *trim boxes* when you set up the cutting room. Most film, however, should be as accessible as possible until the editing is finalized and thus should not be closed up in the boxes but placed instead in open boxes, which can be lined up on your *reel racks.* In addition, should your cutting room have *wooden racks* attached to the walls they are ideal for filing your trim rolls as well as your daily rolls and rolls of cut sequences so long as you keep them separate.

The reel racks are designed to hold reels, but that is the only space available for the trim boxes. Whether you place the boxes on end as in Figure 11–2 or lay them flat, they will perch precariously and the slightest nudge on one box can send it and other boxes scattering to the floor. This can be prevented by placing planks on the racks to simulate shelving thus providing the boxes with support. Of course, you have to remove the plank if you want to use a rack for reels.

Wooden racks attached to the walls offer both additional space and convenience for film rolls. The racks are similar to the shelves on your bench, specially designed

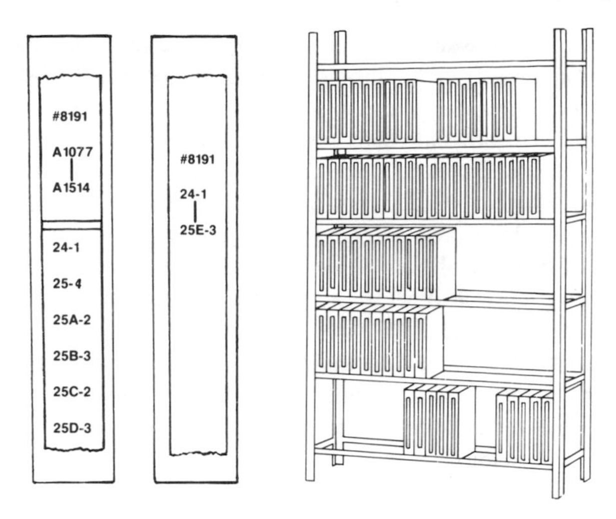

Figure 11–2. Trim boxes.

to hold rolls, but they are constructed of wood instead of metal. Wall racks can help you delay filing your trims into trim boxes until a project is completed.

Your editor will depend on you to be as organized with the filing of film as you should be with the paperwork and supplies. You should try to be so efficient that you can find one frame of film as quickly as you can a reel of film.

You will now have several kinds of film to keep separated and organized:

1. trim rolls,
2. dailies being assembled,
3. dailies broken down, and
4. cut sequences.

Wherever you feel it necessary place identifying strips of labeled tape on your shelves or reel racks so that both you and your editor will know where each kind of film should be.

There are two methods of filing trims that you have wound up from the trim bin:

1. *by scene number,* in numerical/alphabetical order left to right, top to bottom on your shelves
2. *by code number,* in alphabetical/numerical order left to right, top to bottom.

If you are using boxes, each box should be labeled to identify the production number and the scenes or code numbers included inside. Even if they are filed by code number, the scenes involved should be listed (see Figure 11–2). Remember those *outs* (uncut takes) you put aside on your bench? Do not neglect to incorporate them into the proper filing.

If you do not have a wall rack, you must resort to the trim boxes. Leave them open and laid flat on the reel racks. Instead of placing the rolls flat in the boxes, stand them on end just as though they were on the racks or shelves. By so doing, the tabs identifying the rolls will be read more easily and found more quickly.

If you have track and picture dailies broken down to be assembled in numerical scene sequence (see Chapter 8), they should be kept separate from trims as should completed dailies broken down and awaiting editing. Cut sequences should also have separate locations for their rolls and reels and should be filed in scene order.

Other film that must be filed are music tracks, gags, short trims, and *goodies.* Except for the short trims these are usually filed in closed trim boxes and placed conveniently within the editor's reach. Most editors leave the short trim box open.

Music tracks are those production tracks in music takes that are outs, and *gags* are material collected for an anticipated gag reel for a cast party.

Short trims, as previously mentioned, are generally any picture or track trim less than a foot in length. A trim could be as short as one frame. Such trims, if they were hung with other trims in the trim bin, could easily drop loose and be lost. The editor should paper-clip picture and track together and, if neither has a code number, he should identify the trim with the nearest code number with the white grease pencil. Some editors are very careless and neglect to do this, and then it becomes a *detective* job for the assistant usually at a most inopportune time, when that trim is urgently needed. This can be prevented if you check the short trims daily and make certain all are identified.

On projects longer than a series episode, the filing of short trims should be further extended. The assistant occasionally empties the short trims box and files the short trims according to a system agreed upon by his editor. One system is to set up a special short trim file, identifying the boxes as you have your trim file, A 1000–A 3000, A 4000–A 7000, A 8000–A 0000, AA 1000–A 3000, and so forth or Scenes 1–10, Scenes 11–19, and so on. A similar system can be used by placing the short trims into marked envelopes rather than the boxes. I prefer a slightly different system that involves putting each short trim into an envelope that is attached to the inside cover of the corresponding trim box thereby eliminating any sort of extra box file.

Goodies, in our context, refers to any important dialogue or good sound that the editor is unable to cut into the picture because of even more important dialogue or sound occurring at the same time. Initially, the editor should identify each piece with scene and description, such as *Scene 10: Judge bangs gavel,* or *Scene 42: "Hello, there. Anyone home?"* The goodies are eventually given to sound effects (see Chapter 13), and it is then the assistant's responsibility to mark each trim with a specific reel number and footage placement.

AS DAILIES CONTINUE

Shooting and dailies continue and you repeat the same processes as have already been described. Of course the scenes will be different each day, there may be different problems, and the amount of film to be worked on will vary. But your routine for the next five or six days of dailies, depending on whether it is a six or seven-day shoot, will be basically the same as on the first day of dailies with some additional duties required by the editor.

SOUND TRANSFER ORDERS

You will undoubtedly order various sound takes during this period. Your editor may request a *reprint* (another print) of a sound take because it contains ambience (atmosphere background) or background effects that he wishes to add to the picture, or a reprint will be ordered when a track is badly damaged.

Also, the editor may ask you to order a track *outtake,* a take not originally circled. This may be done to seek better line delivery or better background sound. The most common situation that calls for ordering a track outtake is when before you pop your tracks, you have determined that a sound take accidentally was not circled and should have been printed (see Chapter 8).

The sound department of your studio usually has its own transfer order forms. Some of these forms may appear very complicated. Any questions you have concerning these forms can be answered by the sound transfer man or by an experienced assistant. The simplest approach is to write out on the form specifically what you want. For example when you are ordering a reprint, write on the form, "Please make track reprint of 75E-3." When ordering an outtake, write "Please make transfer from outtakes of 37-1." If you phone in an order, plainly mark the form as a *confirmation order* and give it to the transfer person when you pick up the order.

Always include the following information on these sound transfer orders:

1. series title,
2. episode production number,
3. scene (or slate) take number,
4. date shot,
5. the number of prints requested, and
6. the ¼-inch sound roll number that you can obtain from the sound reports (Figures 8-2 and 8-3) in the case of an outtake.

The number of copies required for any order—whether it be for sound, picture, optical, stock, or so forth—varies from one company to another. You will also have to learn who should be sent copies and you will, of course, keep one copy for yourself. A suggested procedure is to keep your copies on clipboards hanging on the wall near your desk. When the orders are received, note the date received on each order, and file the orders in your order notebook.

Whenever it is feasible, all tracks should be coded. Reprints should be spot coded exactly as was the original. Outtakes have to be assigned new code numbers and added to the code sheet for the date of shooting. If for example you were to order 25J-2, you could have it spot coded *A 3396,* which is the end code number of reel 3 (see Figure 9-4). Spot coding it *A 2861* at the end of reel 2 would place this track closer to 25J-4 but, according to the log in the lined script (Figure 10-4), the take runs 2:48, or 252 feet, which means that the code numbers would overlap into scene 25L–3 in reel 3. You must never duplicate code numbers on two different takes.

If your editor does not want to wait for the track to be coded, you should try to identify the outtake wherever he has used it in the picture by writing the scene and take number at the head of the cut with your red pen if he has not done so. All of these efforts to identify the sound may be of future benefit to you and the editor and will be greatly appreciated by the sound editors.

LAB ORDERS FOR PICTURE

You will use an order form of the lab servicing your production. As with sound always include on your order the series title, production number, scene (or slate) take number, date of shooting, number of prints requested, the negative roll number, and whenever possible the negative *key number.*

When you order a reprint, simply write "Please make color reprint." Reprints should be spot coded exactly as was the original print. If a reprint is being used to replace a damaged print, the original should be replaced throughout the cut sequence and in the trim roll and then discarded. If the reprint is for use a second time, it is important that you red line it as though it were for stock (see Chapter 10).

When you are requesting an outtake, write on the order "Please make color print from *B* negative." You will be unable to supply a key number, but it will not be difficult for you to include the roll number. If for example you were ordering 25–3, you would check the lab report (Figure 8–10) containing the printed take (25–4). Obviously, the outtake should be filed under that same negative roll number 147. Picture outtakes must be spot coded using the same procedure as described for sound outtakes. Whenever you are ordering a picture outtake that has sound, always order the track as well and sync and code them both.

KEY NUMBERS

Key numbers are the film manufacturer's numbering printed on the negative stock and are referred to as the negative, or *original,* key numbers. When a negative is developed and a print is made, the identical numbers are printed through on the positive print so *negative cutters* (see Chapter 16) can use these key numbers to match your cut (edited) picture precisely. Whenever you order anything involving the picture from a lab or optical house, you must use key numbers for identification. The exception, as has been mentioned before, is that you do not have the key number for an outtake. Code numbers are used for identification only in the editing process in the cutting room and by the sound and music editors.

Looking at your film emulsion heads up you can find the key number on the left side. It is visible through the black-lined edge of the film. The numbers consistently increase one number from heads to tails every sixteen frames or one foot as do your code numbers on the right, clear side of the film. There are two types of negative principally used in Hollywood, *Eastman* and *Fuji.* Each has a distinctive key number and their placement on the film is quite different.

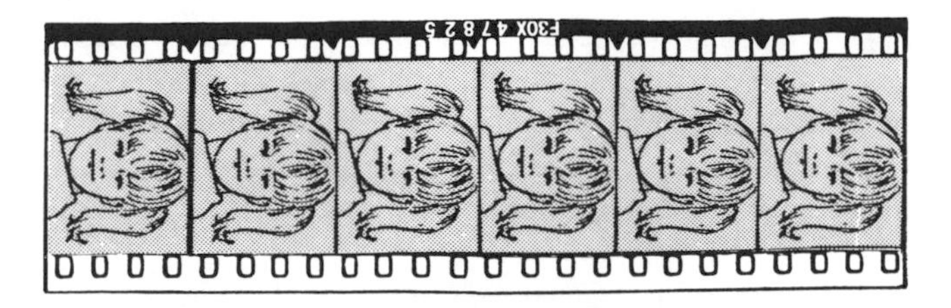

Figure 11–3. Eastman key numbers.

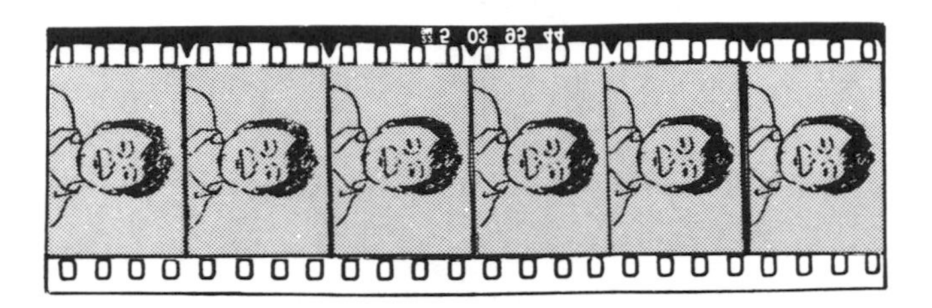

Figure 11–4. Fuji key numbers.

Eastman original key numbers begin with a letter usually followed by two numerals, the letter *X,* and five more numerals—e.g., F30X47825 (see Figure 11–3).

The Fuji original key number is easily distinguished since it begins with a fraction consisting of two numerals over a number and a letter followed by a single numeral and then three pairs of numerals: 22/9M 5 03 95 44 (see Figure 11–4).

FILM CLIPS

A request for a *film clip* could mean anything from three frames to three minutes of film. Longer film clips are most often portions of a scene usually shown on television in connection with a performer's interview to publicize, in the case of episodic TV, the series. Your editor, after conferring with the director and producer, selects the section to be shown. You remove the picture and track from the cut roll or reel, replacing both with leader, and retaining the sync. Mark on the leader information about the temporary use of the film. Also, attach head and end leaders on the film clip to identify the material. Have it delivered as you have been instructed. If instead of the film a film-to-tape transfer has to be sent, you would probably follow the same procedure as you would when transferring dailies to videotape as has been discussed in Chapter 10.

Shorter clips of three to six frames are often requested by the cameraman, costumer, or prop man for matching purposes when a *retake,* or *pickup,* scene or an *insert* is to be shot (see Chapter 12). These clips should be removed, if possible, either from the end of the take or immediately after the slate to avoid the body of the scene, which the editor might need to edit. Sound is not given with these short clips. When you remove a clip, replace it with an equal amount of frames of black leader. If you have not been supplied with *black frame leader,* use the black leader, unframed or framed, the lab generally includes at the heads and tails of your daily rolls. Although black leader is preferable, yellow or gray is sometimes used for this purpose. On a small piece of masking tape attached to the edge of the leader, indicate with a red or black marker the date and the borrower of the clip. You can use a *scribe,* an all-metal pencil-shaped instrument, on black leader. The sharp metal point permits you to scribe, or write, on the film.

You must obviously have your publicity film clip returned, and you should also try to get back the short clips. It helps to remind crew members to return them by attaching a note to the clip with your name, cutting room number, and phone number. Keep a separate record of the clips given out, to whom, and the date on which they were loaned. Also *scribe* on each clip some identification such as the nearest code number that will help you locate its original position when it is returned.

TAKE-UP MOVIOLA

Some pick-up or retake scenes may be so involved that the director may ask that the previously shot, related material in daily form or in cut sequence be brought to the sound stage to be viewed before the shooting. Arrangements must be made through the post-production supervisor's office for a *take-up moviola* (Figure 11–5) to be delivered to the stage, and you might be asked to run the film.

There is a considerable difference between the operation of the upright moviola most commonly used by a film editor, sometimes called a *cutter moviola* (see Figure 9–2), and a take-up moviola (Figure 11–5) that has attached arms to hold reels so that, as the film is run, the film is wound up or can be rewound as on a projector. The picture and track reels are placed on the lower spindles. The empty take-up reels are placed on the upper spindles. The film is threaded from the lower reels through the rollers and sprockets and up through the picture and track gates and is then taken up by the

Figure 11–5. Take-up moviola.

upper reels. It is important to provide a loop (about eight frames of extra footage) in the picture before it enters the gate to relieve the tension on the film. Always have the film run down to the section the director or a crew member wishes to see before the viewing so that no unnecessary time is spent locating the scene thus delaying the shooting.

Take-up moviolas are principally used by sound and music editors so if those editors are employed where you work it behooves you to find someone who can take the time to instruct you how to thread up the film and then to practice running it in advance. It would be most embarrassing if you were to enter a sound stage in the midst of cast and crew and discover you did not know, or had forgotten, how to thread up the film. You can avoid this kind of situation by reviewing the procedure beforehand.

ACADEMY LEADERS

A good assistant takes pride in trying to anticipate the editor's needs and in having material ready before he even suggests getting it. Just as you prepared yourself for the first day of dailies by making up your daily leaders the previous day, so also sometime before shooting is completed when you have some spare time, you should prepare for the time when your editor will edit all the scenes into story sequence. *Academy leaders* are needed for the beginnings of each reel; and *commercial banners,* for placement between the acts.

Academy leaders are industry standard track and picture, head and tail, projection leaders printed by a lab for release prints and also used on edited work prints. They are 12 feet from the start mark to the end of the leader where picture commences with decreasing numerals on the film beginning with the 12-foot start. Some *academies* are marked with seconds, instead of footage, beginning with *8* at the start. They are still 12-foot academies because 8 seconds = 12 feet of film.

Some companies do not provide you with track academy, and in this case you have to use ordinary leader for the track matching it with the picture. In any case on both leaders emphasize the start mark by placing a piece of masking tape across it and marking the tape with a large *X,* black on picture and red on track, and identifying it as the "——— 12-foot START ———."

You also have to provide a *head pop* at the three-foot, or two-second, mark of the track academy leaders. At this point cut out the one-frame mark and splice in one frame of *sound tone.* This sound tone, or *pop* (not to be confused with the clapstick pop), can be salvaged from the heads of your daily tracks. The pops may also be supplied to you on removable tape that is frame length and can be taped across the sound side of the frame thus eliminating the necessity of any splicing.

For a one-hour series, six or seven academies should be prepared for the first cut, thus anticipating that most first cuts are considerably over footage and require six or seven reels even though the film will be finalized on five reels. Splice four or five feet of blank leader to the heads of the academies with the following information on the leader: episode production number, series title, reels 1 through 6 or 7, *Picture* and *Track,* and *Head.*

Also prepare temporary ten-foot end leaders for each of the reels with similar identification, except instead of *Head,* use *End* or *Tail.* Before the final cut you will have to add ten more feet to these leaders for a total of at least twenty feet (see Chapter 13).

COMMERCIAL BANNERS

A one-hour show is interrupted for commercials three or four times depending on the *network format requirements* for that particular show. For our example scenario, let's assume that your format (see Figure 13–3 and other material in Chapter 13) requires four acts separated by format material including commercial banners. When the editor combines all the sequences he has edited into story sequence after shooting is complete, he will want to include, temporarily, only the commercial banners at the *act breaks* so that during their screenings the director and producer will understand where one act ends and another begins.

Commercial banners, or commercial *leaders,* like the missing banners discussed in Chapter 6, may be provided for you with several different titles. The plain titles against a black background usually read *commercial* or *place commercial here.* They are ordered from the lab by the post-production supervisor and come in 1000-foot rolls. The assistant cuts off about 150 feet, which should last him through the television season. The format requirements for the lengths of the commercial leaders vary from one foot to three feet. Also required by the formats are one to three feet of black leader attached to the heads and tails of these banners.

Some companies will have the lab print up three-foot commercial leaders, each one with three feet of black already at the head and tail. You will have to make any length adjustments required by your project's specific format.

Refer again to the format in Figure 13–3. For the first act break between acts I and II you need a one-foot banner (see 10) preceded and followed by three feet of black leader, between acts II and III another one-foot banner (14) is required preceded by three feet of black and followed by two feet of black; and between acts III and IV the format requests that the banner (23) be one foot and the black leader three feet before and two feet after.

Remember that picture and track must always be kept in sync and that therefore you should not forget to include leader of equal length for track with each commercial banner, including the black leader. The other commercial banners and material required by the format are not inserted until before the final screening of the picture.

EXCEPTIONS

Kem Rolls

Editors who use a horizontal, or flatbed, editing machine (Figure 9–5) employ a variety of filing systems for their film. Most of them begin editing with *Kem rolls* as discussed in Chapter 10. In order to maintain sync in these rolls, a constant *reconstituting,* or splicing in of leader, is required whenever an unequal amount of picture or track has been removed and used. As trims are made during editing changes, some editors will revert to a moviola system and file them into trim boxes thus eliminating further reconstituting.

Location Shooting

When principal photography is shot far from Hollywood, editing facilities may be set up on location. Your responsibilities may increase, and there will be no other assistant or apprentice for any kind of assistance. If you are given an Acmade, you will have to do your own coding. Without studio facilities or a post-production supervisor—although there will be a production manager, he usually does not function in editorial matters as thoroughly as does a post-production supervisor—you will be expected to follow through with various matters relating to the transportation of film or videotapes between location and studio, the maintenance of equipment and supplies, and the expediting of lab and sound dailies and orders.

Even when you remain at the studio, location shooting can cause you additional work in transporting material. In addition when film dailies instead of videotape transfers have to be sent to and from the shooting site, it will result in delays in editing. Also, considerable time can be spent trying to establish communication with members of the crew when problems arise.

DAY SEVEN

It is the end of your seventh day as an assistant editor. The first episode of your series completed shooting in six days and you have broken down your last day of dailies. On this day the crew began shooting on the second episode, which will be edited by the second *team.*

All of your dailies are recorded, coded, broken down, and available to your editor. All your paper work is properly filed. Academy leaders and commercial banners are lined up in individual rolls for him. It will be at least another twelve working days before you will begin receiving dailies on your next episode. Your time can now be devoted exclusively to assisting your editor in preparation for the *director's,* or *first, cut*—the first screening of your picture.

CHAPTER 12

THE FIRST CUT

There is a commonly accepted aphorism that the theatre is an actor's medium while motion pictures are a director's medium. Why? Because in the theatre the actors are in control once the curtain rises while in film the director is presumably in control of what will be viewed. This is certainly a misconception when it is applied to episodic series where, although the director is certainly important, the *producer* has most of the creative control.

In episodic television the director is entitled to a *director's cut* normally within six days after the last day of shooting. Before screening the assembled picture for her or anyone else, the editor should be given the time to run an *editor's cut* for herself and subsequently to make any necessary changes. Except for occasional suggestions from the director or producer during dailies, the director's cut is solely the editor's version. Since the director's changes are the first changes to the editor's original version, this screening is also known as the *first cut.*

The only limitation imposed on the editor as she edits her original version is to try to tell the story as she believes the director intended and as indicated by the material that was shot. This requires technical and creative expertise, and the editor and assistant work diligently to make the first cut as good as possible. Therefore it is understandable that any reference to a first cut as a *rough cut* or an *assembly* is denigrating and offensive to editorial personnel.

At the first cut the director will ask the editor to make certain changes, additions, or deletions. Following this screening, however, at the second screening, which the director may or may not attend, the producer will take over. What the television audience ultimately views will be the result of the producer's editorial supervision: *Episodic film is a producer's medium.*

Although the producer may ultimately make the editing decisions, the specific implementation of those decisions can usually be done in several ways and often may diversely affect related areas that the editor must also re-edit. Therefore in addition to keeping within the required time schedule the editor must accomplish this in the best way possible consistent with the direction, the performances, and the spirit of the script.

PRINCIPAL ASSISTANT DUTIES

Once shooting is completed, the script supervisor delivers the original lined script. Until this point the editor has been receiving photocopies. When all dailies are broken down, she either continues cutting the remaining sequences or simultaneously cuts these sequences as she *builds* the picture beginning with reel 1, scene 1 and incorporating academy leaders, cut sequences, and the commercial banners as she progresses.

Unencumbered by dailies, you can now concentrate your attention and energy on assisting your editor. Your major duties in preparing for the first screening of the cut picture are listed below. You will also have most of these same responsibilities for future screenings.

1. Have the head academy leaders and commercials ready to give the editor when she asks for them before the first screening.
2. Give your editor the cut sequences in the proper order as she requires them.
3. Line up any uncut dailies for your editor as needed.
4. Get trims for your editor as requested.
5. File the trims and keep your editor provided with an empty trim bin.
6. Order any tracks, picture, reprints, stock, and so forth she requests and make certain any material ordered is received in reasonable time.
7. Assist your editor in the selection and insertion of temporary sound effects and music.
8. Splice the reels including temporary end leaders.
9. As you splice check for any nicks or tears and repair them, and clean off any unnecessary grease pencil marks.
10. Check for sync and framing.
11. Measure the *action* footage.
12. Keep a *screening record.*
13. Indicate changeover cues.

MORE ABOUT TRIMS

As she builds the picture, your editor will be editing some dailies not yet cut. These trims will be filed accordingly (see Chapter 11). Your editor, however, may want to make some changes in the edited material that will result in miscellaneous trims. Although some editors may have an organized system for separating these trims, most will simply hang them in the bin in bunches and rely on you to separate and properly file them.

There are two ways of filing these miscellaneous trims, depending on your editor's preference. You can either roll up the trims by groups of assigned scene numbers or file each trim in its appropriate trim roll.

In using the first method you make a tab for each group of trims. All the tabs can be identified as *Miscellaneous Trims*. Each tab is assigned a number of scenes, such as scenes 1–10 then scenes 11–20. You will need a separate file of miscellaneous trim boxes for them. Situate the boxes on your racks usually at the end of your major trims. This is a simple and fast system that is adequate for projects of short duration like an episode of a series.

Longer projects require the second, more precise system—filing each trim in its appropriate trim roll. How will you know in which trim roll a particular trim belongs? Check the code number on the trim and, if your trims are filed by code numbers (see Chapter 11), go directly to the trim box holding that code number. If, however, you are using the scene system, you have to check the code number in your code book for the corresponding scene number and go to the box that contains the scene. Then you unwind the trim roll into a bin, insert the trim in the proper place, rewind the trim roll, replace it in the box, and return the box to the rack.

The editor and assistant may decide to use *both* methods to accommodate trims that contain several cuts within the same trim (see Chapter 13).

An organized and efficiently maintained filing system will save considerable time in the cutting room and will eliminate the danger of ever losing a piece of film. Your editor may need a roll trim to replace a cut or to add to a cut. Your editor should only need to call out a code number for you to be able to find the appropriate film quickly. Referring to your hypothetical dailies, if your editor asks for *A 3279,* what would you give her? It should be scene 25M–2. What if she asks for *A 1067?* Give her scene 39B–1. Refer back to Figures 9–3 and 9–4 and test yourself.

What would you do if you were using the code number filing system and your editor, instead of asking for a trim by code number, asked for it by scene/take number? How would you locate the correct trim box when the boxes are filed by code numbers? Having worked with the trim boxes within the past few days, you might recall the general location of the scene, and by checking the scene identifications on the boxes you should ordinarily find the trim roll quickly.

If this method fails, however, how can you locate the trim roll specifically? Find the scene in your code book by flipping through the pages until you find the scene and take, which lists the code number. But this could be time consuming. Your editor can save you some time by giving you the date the scene was shot, which she can easily find in the script log (see Figure 10–4). Since your code sheets are in date-shot order, you can then quickly find the take and its code number and this information will lead you to the exact trim box and the trim roll.

REPAIRING FILM

It is extremely important for you to identify nicks and other defects in the film and repair them as best you can. One small nick on a sprocket could result in irreparable damage to several feet of film and interrupt an important screening. Even a bent section of film can create problems in projection. Defects on picture should be repaired using clear tape on both the cell and emulsion sides. When you are making any sort of repairs across the full width of the picture that require more than the standard two-sprocket splice, make certain that both ends of the tape end at a frame line that separates each picture frame to reduce the visibility of the tape on the screen. A bent section of film should be reinforced with tape usually only on the cell side to prevent a future tear.

Nicks, if they are restricted to sprockets and the edges of the film, may be repaired without covering the full width of the film. For example if two sprockets next to each other are nicked, use a scissors to cut a frame of clear tape. With the help of the center pins of the butt splicer, place only the tape sprockets over the damaged film sprockets so that you are not taping over any portion of the picture frame. Now turn the film over and bend the tape so that the opposite tape sprockets cover the same sprockets on the reverse side of the film. Cut off the bent portion of tape even with the edge of the film. The nick is now repaired without taping over the entire picture frame (see Figure 12–1).

No matter how carefully you may have repaired severe tears across the picture frame whether for several frames or for a couple of feet it may be momentarily disturbing in a screening. Even worse is the possibility that the damaged area might tear again during a screening. You should try to avoid this by phoning the lab immediately after you discover the tear and initiating a confirmation order for a *rush* reprint so that you might replace the damaged cut before the next running.

Sprocket nicks and tears on track are repaired with white tape on the cell side and, because you need not be concerned with seeing the tape, the taping can be done across the full width of the film. Even though a damaged track may not be affecting any important dialogue or other sound if it is an extensive tear, get a reprint quickly to replace the damaged cut.

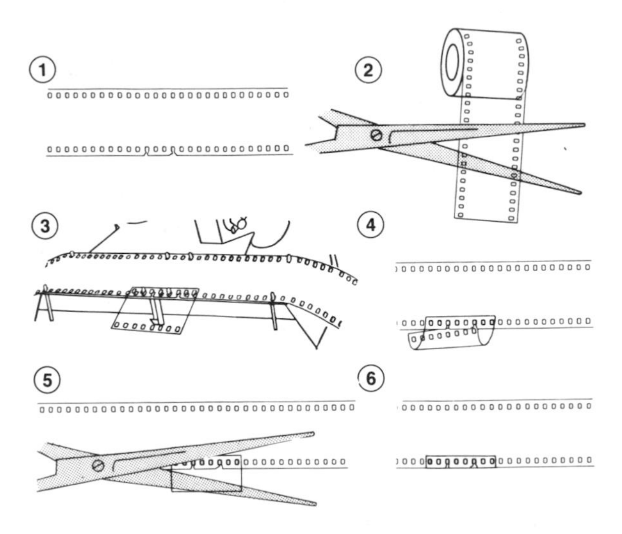

Figure 12–1. Repairing picture sprockets.

REMOVING EDITOR'S MARKS

During editing the editor uses a white grease pencil to make marks on the film. Some of these, such as optical marks (see Chapter 14) or the editor's sync marks that are generally on the edge of the film where they can't be seen during a screening, must be left intact. But grease pencil marks left unintentionally on the picture by the editor should be cleaned off with a Webril wipe. These marks are usually those made by the editor to indicate the beginning or end of a cut she was considering and then accidentally left on the film after she made a different decision.

If your are uncertain whether to leave or remove a mark, always ask your editor. In a short time you will become familiar with all of your editor's symbols and will instantly know which marks to leave and which to clean off.

CHECKING FOR SYNC

It is the assistant's duty to verify that picture and track are in sync. The editor will depend on you to check sync and make any necessary corrections or to ask her to make them. Anything that disrupts a screening such as the film being out of sync will interfere with a director's, or producer's, honest reaction to the editing and the visual story being told.

After each reel has been *built* or reedited by your editor and spliced by either the editor or you, you will check the reel for proper sync. At the same time you will also verify that the film is in frame, measure the picture's footage, and mark or confirm the change-over cues (guides to the projectionist for switching reels). These simultaneous tasks are discussed below. Finally *your* last check of the reels will immediately precede the screenings for the director and the producer so it is obvious how responsible you must be.

To check the sync (and to perform the other simultaneous tasks) you will need to run both picture and track through the synchronizer. Since you just spliced the film that the editor gave you tails-out, the reels are now heads-out, and thus you will be winding down from heads to tails. Put the reels on the left rewind and run the film through the synchronizer and begin with lining up the picture and track start marks. As you wind down the reels, make certain that like code numbers are aligned. If identical picture and sound code numbers are exactly within the corresponding frame and even with each other when the two films are viewed side by side, they are in sync.

A variance of one or two sprockets is not noticeable in a running, and many assistants, particularly if they are racing the clock, will not take the time to correct this sort of variance. However, if there is an additional discrepancy of a few more sprockets further on in the reel, it could compound the initial error and the film could become more than a frame out of sync. This could seriously affect the screening. Therefore to avoid this danger it is suggested that whenever even *one* sprocket variance is involved it should be corrected. The correction can usually be made by an addition or deletion in the track at the splice at the head of the cut where the out of sync first occurs.

When an addition is required you should retrieve from the trims the sprockets or frames of track that the editor originally omitted, using the code numbers on the film as a reference and making certain that the sprockets or frames do not contain any undesirable sound such as the end of a word or a noise, before splicing the addition in. Do the splicing on the *left* side of the synchronizer to maintain sync ahead of the splice and to verify the adjustment within the out-of-sync cut. Should the sprockets from the trim roll contain undesirable sound, get substitute sprockets (*fill*) from the same trim at the head or tail of the take. The ambience, or background, sound will then be consistent.

Should you have to delete track, make your adjustment at the same splice at the head of the cut. Get the trim roll and splice the deletions to the proper trim. First, check whether the sprockets of track contain any important sounds that belong there. If so, you may be able to eliminate the necessary track from the end of the preceding cut, or it may be an indication that track should not be deleted but rather that picture should be added. This usually occurs when the film is out of sync by a frame or more.

However, do not be too hasty in judging a film to be out of sync just because the code numbers of track and picture not only do not match but are completely dissimilar. This occurs when a particular track *overlaps* other angles. For example, consider a scene in which a speaker addresses an audience. The first cut is of the speaker and the track and picture code numbers match exactly. Then, continuing with the same track, there are several cuts of different individuals in the audience listening and reacting to the speaker. Obviously, none of these picture code numbers match the speaker's track. Yet there is no sync problem that needs correction because when the film finally cuts back to the speaker, the track and picture are in perfect sync. A *prelap* is when a speaker's track is over other cuts *before* the speaker is seen. Always check overlapping and prelapping tracks and picture thoroughly before concluding you have an out-of-sync situation.

It is always prudent to consult with your editor before adding or deleting any *picture* in order to correct an out-of-sync. She will probably want to make the correction herself anyway or, at least, give you specific instructions about what to do. When you are adding picture, as with track the nearest code number is used to get the necessary frames from the trims. When you are deleting picture, also refer to the nearest code number so that you can splice the extra frames to the proper trim.

CHECKING FOR OUT OF FRAME

Just as you were cautioned about being in frame when you were syncing dailies (see Chapter 9), so must you verify that the cut picture is not out of frame anywhere. As you roll down a reel checking the sync, periodically affirm that the picture frame lines are lined up with the framer or frame line notches on the synchronizer (Figure 4–5). An out of frame usually occurs when the editor misjudges the frame line and lops off one less sprocket than she should have or adds a sprocket from the next frame. You must either find that missing sprocket in the trims and splice it in or, in the second instance, splice off the extra sprocket and add it to the appropriate trim.

MEASURING THE PICTURE

As you check each reel for sync and framing, you will simultaneously measure the picture. Your objective is to measure *action footage* only. Action footage is only that film that *tells the story.* After you have lined up picture and track at the 12-foot start mark to prepare to check sync, since the academy leaders are not included in the action footage, roll down in your synchronizer to the first frame of picture and line that frame up with the first frame of the *framer* (see Figure 4–5). At this point set the footage counter at 0000 so that you will begin measurement at *0000 + 01* at the first frame of picture.

Never measure the commercial banner, and unless the *network format* (see Figure 13–3) indicates otherwise, do not measure the black leader before and after the commercial. In whatever reels you find such leaders, which signify an act break, record the footage at the very last frame of action preceding the break and then reset the footage at 0000 + 01 at the first frame of picture action that follows. For each reel or part of a reel you measure, take down the footages indicated on your sync machine counter plus the number of frames on the framer.

SCREENING RECORD

Before every screening you as the assistant should keep a record of information that might be required on your project. This is particularly vital on any kind of television show where network footage requirements have a crucial effect.

Companies do not usually provide the assistant with a form for this kind of record so assistants have to make up their own form. The assistant's screening record shown as Figure 12–2 is a suggested form. It contains the kind of information regarding the status of your picture at the time of the screening that might possibly be asked of you by the editor, the director, the producer, or the post-production supervisor. As you measure footage and check sync on the picture and simultaneously record the footages, also note the other pertinent information.

In the first column of Figure 12–2 the cut or screening (1st cut, 2nd cut, and so forth), the date, and for whom the picture was run is recorded. In the figure the 1st, or director's, cut was screened for the director, Carla Allen, on October 18. In the second column enter the scene number that begins each reel. Reel 1, it is assumed, generally begins with scene 1, the opening scene.

But how can you quickly determine the beginning scenes of the other reels? You check the first code number on each reel, of course. In Figure 12–2 reel 3 of the 1st cut begins with scene 25, but suppose instead that the first code number on the reel is *A 1394,* and by checking the code sheet (Figure 9–3) you learn that the reel begins with a cut from scene 25C. However, scene 25 is a long scene and merely noting *25C* does not indicate precisely where in scene 25 the reel begins.

Therefore in the third column carry this information one step further by noting a description of the beginning action or dialogue. By this time you should be quite familiar with the episode, and a quick glance at the film will enable you to give a brief description of the action. Should this beginning action be dialogue simply place the

DATE FOR WHOM	BEG. SC. #	DESCRIPTION	REEL	REEL FTGS	BREAK FTGS	ACT	ACT FTGS
10/18 1st CUT Carla Allen (director)	1	Estab shot, San Diego	1	862+10	862+10	I	1259+15
	13	Stretcher from hall-way of Marshall home	2	881+03	397+05 483+14		
	25	Enter Kirkwood/Ryan/Bellamy enter Marsh. apt.	3	889+03	889+03	II	1507+10
	38	Ext. Golman's Apt bldg. Kirk/Ryan drive up	4	851+04	134+09 716+11	III	1498+11
	50B	Kirkwood's office - Phyffe enters	5	927+07	782+0 145+07		
	61C	Dets question Fern- "What was his motive?"	6	756+12	756+12	IV	1573+12
	70A	Ryan enters elevator	7	671+09	671+09		
		TOTALS—		5840+0	5840+0		5840+0
		Ftg. Over:		1653+0		Time Over:	18:22
		Ftg. Under:				Time Under:	
10/20 2nd CUT Ned Forbes (Producer) Carla Allen Tom Roberts (assoc. Prod) J. Crawford (P.P.S.) Nancy Smith (Prod's Secty)	1	Estab. shot San Diego	1	912+03	912+03	I	1242+01
	15D	Charlie exits kitchen	2	947+06	329+14 617+08		
	26G	Kirkwood office — Hancock exits	3	956+03	861+11 94+08	II	1479+03
	45	Dets car L-R into police station	4	938+04	938+04	III	1465+12
	59	Ryan handcuffs Mendoza	5	951+12	433+0 518+12		
	65C	Group at restaurant table - Kirkwood; "When did you see her?"	6	894+04	894+04	IV	1413+0
		TOTALS—		5600+0	5600+0		5600+0
		Ftg. Over:		1413+0		Time Over:	15:42
		Ftg. Under: —				Time Under: —	

Figure 12–2. Assistant's screening record.

sound head of your sync machine on the track, turn on your amplifier, and write down the first line of dialogue.

Why should you need the information in columns 2 and 3? Suppose a director or producer suddenly requests a review of a particular scene on the moviola, and the editor asks you to wind down the picture and track to the beginning of the scene in the synchronizer on his bench. After leafing through the script and verifying the scene number, how would you know which of six or seven reels to go to without the above information? The screening record shown as Figure 12–2 pinpoints exactly where each reel begins so that you can quickly determine the reel in which the scene begins, even eliminating the necessity of going through the script in most cases.

Reel footages (REEL FTGS) in the chart are the *total* action footages of each reel. The *break footages* (BREAK FTGS) column also lists the total footage of a reel except that it divides the footage into two parts if the footage on the reel is separated by a commercial break. Thus in Figure 12–2 the break footages of reels 1, 3, 6, and 7 of the 1st cut, for example, are simply their total footages because these reels have no commercial breaks. Reel 2, however, has a break at the 397 feet plus 5 frames point, and thus the two sections of footage preceding and following the break are listed separately. Similarly, reels 4 and 5 have two sections of footage recorded because they both have commercial breaks.

These commercial breaks signify the end of an act and the beginning of the next act. Thus, reel 1 is uninterrupted. It contains material that is entirely part of act I. Act I ends in the middle of reel 2, at 397 feet + 05. The last two columns of the screening record list the acts and act footages. To obtain the act footages add together parts of different reels. For example, the footage for act I in the 1st cut is 862 feet + 10 (all of reel 1) plus 397 feet + 05 (reel 2 up to the first break), which totals 1259 feet + 15.

Note that all footages are recorded as *feet plus frames*. When you are adding footages, do not neglect to add frames and feet when your addition is 16 or more frames. For example, if the total of act I was 1259 feet + 16 (instead of +15 as shown in the last column) the correct answer would be *1260 feet + 0*. Similarly 1259 feet + 19 would be *1260 feet + 03*.

The totals of the three footage columns should correspond exactly. If there is any variance between them, an error has been made somewhere and should be corrected.

If a commercial occurs at the end of a reel, indicate this on the screening record. When this happens, most editors prefer to split the break material by placing the commercial, and the black leader preceding it, at the end of the reel. The next reel begins with the remaining black leader.

Ideally, the post-production supervisor will have a network format for the series before the 1st cut so that you will be advised of the footage requirements. The editor across the hall who is cutting a different series may already have a format but you cannot use hers. Even though it may be for the same network, the format will probably be completely different than the one for your series.

The format tells you exactly how much *action footage* is required. In the case of the hypothetical show, on page 2 of the example format (Figure 13–3B), it directs you to *cut basic show* to *4187* feet, or 46 minutes, 31⅓ seconds. You can determine the current footage status of your show by subtracting the format footage from the total footage of the director's cut. In this case, 5840 feet + 0 less 4187 feet + 0 is 1653 feet + 0, meaning that you are 1653 feet + 0 *over* footage. Although a show is rarely *under* footage at this stage, if you were under footage, you would subtract the cut footage from the format footage to give a negative, rather than a positive, number.

You and your editor will be thinking in terms of footages during the editing process, but the director and producer will probably want to know the amount of *time* that must be deleted. You can calculate this from the excess footage quickly by using the *footage chart* (Figure 8–10).

Example:	footage over–	1653 + 0		
	nearest footage on chart–	1620	= 18:	(minutes)
		33 + 0 =	:22	(seconds)
		Total Time Over =	*18:22*	

Since the screening record form in Figure 12–2 is designed for a half-hour or a one-hour TV show, it would have to be extended for the additional reels of TV movies and miniseries. Theatrical films, which have no commercials or act breaks until they appear on television, would not require the three right-hand columns of the form.

CHANGEOVER CUES

When you arrive at the end of each reel, except of course the last reel, as you are checking sync and measuring, you have to *cue* it so the projectionist can make the proper changeover to the following reel. This will insure that no part of the tail or head leader will be viewed and interfere with the continuous action of the picture.

You create a changeover cue by making a diagonal slash mark with your white grease pencil from screen upper right to screen lower center across four picture frames on the emulsion side of the film. To fully cue a reel you must mark two sets of four consecutive frames, both sets separated from one another but located near the end of the reel.

Again, unfortunately, there is no standard location for the cues. At most companies the first cue is placed at 12 feet plus 1, 2, 3, and 4 frames from the end of the reel and the second cue, at 1 foot plus 1, 2, 3, and 4 frames from the end (see Figure 12–3). Some companies may require a slightly different placement. Ask another assistant to loan you her *template,* a sample reel end on plain leader that is marked with the changeover cues at the proper footages. Make a copy of the template, and use it as your guide.

The next time that you go to a movie, if you watch the right side of the picture, you will see near the end of a reel some small light circles. These are the changeover cues, also called *birds-eyes,* provided by the lab on the composite print. You may recall an episode of the original *Columbo* series in which a knowledge of these changeover cues helped Detective Columbo solve a studio murder.

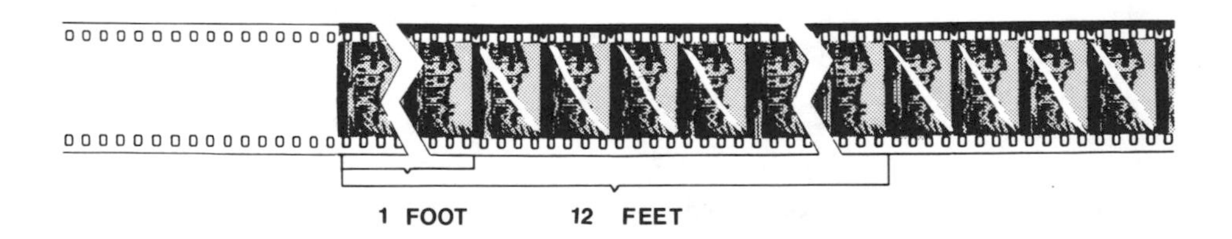

Figure 12–3. Changeover cues.

OTHER ASSISTANT DUTIES

Since you do not have dailies to prepare during this period, you can focus completely on assisting your editor on the first cut. Follow through on your editor's requests as quickly and efficiently as possible. You should easily be able to keep up with her in performing your principal duties. It is rewarding and fun to try to anticipate her needs and have the materials ready even before she asks for them. Volunteer to do anything you can think of that will save her time and contribute to the screening.

You may have other tasks in addition to your principal duties when you are preparing for screenings. These involve opticals, retakes, pick ups, inserts, stock, ambience, sound effects, and music.

Opticals

Opticals are a variety of effects made in the optical department of a lab or an optical *house* with production material. Do not confuse *optical* effects with *special* effects. Examples of special effects would be a great fire as in *The Towering Inferno* or the sinking of a ship as in *The Poseidon Adventure.* Other examples of remarkable special effects are in such films as *Star Wars* and *Raiders of the Lost Ark.*

Optical effects are the transformation of production material into fades, dissolves, superimposures, flopovers, and so forth. Opticals are not usually ordered until the picture is *locked in,* meaning that the picture editing is completed. The term *locked in* is also used to refer to an editor's being restricted to editing a scene or portion of a scene a certain way according to the manner in which it was directed. As used here, however, it means that the picture has been edited to the correct footage; it is *on footage* and has been given final approval.

If an optical is ordered earlier during the editing, its footages might change or it might be eliminated entirely resulting in unnecessary expense. Exceptions to this would be when the effects play such an extraordinarily integral part in the film that it can not be completely edited without the opticals. Another exception would be the case of a necessary, complicated effect that requires approval at the earliest possible screening. Find out who your liaison, or contact, person is in the optical facility to better expedite your rush order. A detailed description of the various kinds of opticals, how to order them, and how to edit them into the picture is given in Chapter 14.

Retakes, Pick ups, and Inserts

If any of these were not shot by the director of your episode, they will probably be shot by the director of the second episode within the first days of that shoot. Advise the assistant on that episode of the material you are expecting so that assistant might watch for it and give it to you as soon as she receives it. Although it should be identified with your episode production number, it will be delivered with the other assistant's dailies. Since it is for your episode, you are responsible for its syncing and preparation for coding.

A *retake* is an angle that had been shot on a previous day and that the director decides to shoot again. The script supervisor will identify it with an *R* preceding the scene number. If, for example, it is scene 39B–1, the over-shoulder shot of Kirkwood and Ryan (see Figure 8–1), you would receive this retake slated as *R39B* with the take number as circled.

Scene 25 (see Figure 8–1) was shot and take 4 ordered printed, but the director wished to reshoot the end of this shot overlapping it into succeeding action. Therefore the scene is identified as 25A, a *pick up* of scene 25. This kind of pick up can also be any section of a shot the director wishes to reshoot to save time as opposed to reshooting an entire scene, i.e., Bellamy's pick up (P.U.), 25E, and Kirkwood's, 25I. Another type of pick up is a take the company neglected to shoot or for some reason could not shoot. In episodic television this pick up would probably be shot later and probably by the director of the following episode. It would be identified, according to your script supervisor's preference, by the next letter past the last letter used in the original shooting or by adding a letter. If, for example, a pick up was shot for scene 39, it could be slated 39C, 39AA, or A39.

An *insert* is a close shot of an object or piece of action that is significant in the story. A typical example of an insert would be a closeup of a newspaper headline or a closeup of a desk drawer as a hand enters the frame and opens the drawer to reveal a pistol. An insert can be identified in the same way as can a pick up. The inserts in your script notes, 25M and 25N, were shot during regular production but inserts are often shot in post-production to save shooting time. In the latter case, they are often *supervised* by an associate producer, the editor or, as on Mission Impossible for example, even by the assistant editor.

Any retake, pick up, or insert being shot after regular production is over will require that you furnish film clips and sometimes it will be necessary for you to run material in the take-up moviola on the set as described in Chapter 11. The script supervisor will send the editor a copy of the lined script showing the additional shot. If the scene is shot after the original lined script has been given to the editor, the script supervisor will borrow the pages involved so that she can include the additional shot on the lined page and the log. This will generally occur only if the material will be shot by the director of the following episode, but the material will still be identified with your episode production number.

Stock

Film shot for your episode to be filed as *stock* was discussed in Chapter 10. On a new series it is generally standard procedure for a crew to go on location to shoot only title and stock material before principal photography on the first episode begins. This footage consists of establishing shots, car runbys, and so forth. This gives the film librarian the beginning of a library for the series so that prints, when they are needed, will be available. If there is no librarian, however, and if you are the first assistant on the series, you will probably be assigned responsibility for the library with help from the other two assistants. This means keeping an updated catalogue and a file of cutting prints.

If the editor needs a certain stock that is not in the studio film library, calls must be initiated to other company and independent libraries. The film, which is for viewing purposes only, is delivered on reels or in rolls, and either you or the librarian will go through the film and select choices for your editor to consider according to your understanding of what is required. Identify the choices by separating those in rolls or *papering off* the selected scenes in the reels, which means folding pieces of paper around the film before and after the section of film that is being considered and marking them *start* and *end* so that they might be easily located. Never use paper clips to secure the papers because the editor may accidentally run past a clip and severely damage the film.

If stock from another library is selected, you will have to call your series' librarian and order a cutting print. Borrowed library film should be returned as soon as is possible. Keep a record of any alternate choices in case the director or producer should have some objection to the film selected and want another choice.

When the cutting print arrives, red line it as you do with production stock and tab it with the name of the library and the can, roll, and key numbers. The trims can be placed in a trim box labeled *Outside stock trims.*

Ambience, Sound Effects, and Music

Some editors over a period of time collect an *effects* (*EFX*) *library* of their own with some commonly needed material that they can use temporarily until the picture is *turned over* to sound effects (see Chapter 13). If your editor does not have a particular effect, you can probably obtain it from your sound effects department.

You may be asked to *fill in,* or cut in, some of these effects. *Ambience* is the pervading sound atmosphere or *room tone* of a particular scene. Assume, for example, that in our hypothetical scene 25 the editor had to cut in leader for *fill* in several places thus causing dropouts in sound. You can probably find enough of that apartment ambience in the clear at the head or tail of the master scene or one of the other takes to make a *loop.* A loop is a section of track that is joined or spliced beginning to end, forming a continuous track for the purpose of rerecording. Make a 35mm mag transfer of the loop for sufficient footage to replace the leader. Your editor might also request *sound effects* for a sequence that was shot MOS (silent) or to provide an effect not in the production track that is an important story point or that cues a specific action.

If you and your editor can find some *music* that fits the desired mood of a scene and will enhance the telling of the story, you should get a 35mm mag transfer of the tape or record and cut it into the work track without disturbing any important dialogue or other necessary production sound.

Why go to all this bother when the music and sound editors will eventually replace all these temporary embellishments? These efforts will add immeasurably to the screenings. By eliminating dropouts in sound or replacing disturbing production tracks, these sections will *come alive* and increase the director's and producer's ability to react clearly to the picture. Also, your contribution to the screenings will now be more than the technical details of having made secure splices, accurately measured footage, and checked sync. Your *creative* work will be in the film enhancing the editing.

The Editor's Cut

As was mentioned at the beginning of this chapter, once the picture is edited and assembled in script sequence the editor is entitled to take the time to run it for herself and make any additional changes she might consider necessary before screening the film for the director. This screening is the *editor's cut.*

Usually the editor screens the editor's cut in the moviola and you assist her in making any further changes as she goes through the picture. If instead the editor first runs the film in a projection room, you should note the changes she will make with your help in the cutting room. This screening also gives her the opportunity to double check any effects, music, stock, or other material you may have cut into the picture.

Should the editor make only a few changes, she may help you keep track of footage lost or gained so that you can make the necessary adjustments in your screening record.

Screening the First Cut

As was recorded on your screening record (Figure 12–2), your picture is on seven reels and you are responsible for the proper and prompt delivery of the reels to the projection booth. Bring with you your lightboard, writing materials, and the lined script. Also, bring the footage measurement and timing as was discussed earlier in this chapter.

Give yourself time before screening the first cut to acquaint yourself with the control panel in the projection room since you may be asked to operate it. Some panels are quite elementary, only an intercom with which to direct the projectionist. Others may have complete controls for starting, stopping, and going forward and backward.

Try to find out before the running what type of screening the director prefers so that you can advise the projectionist accordingly. There are several kinds of screenings:

1. *straight run-through*—without stopping making comments as it runs or afterwards going through the script and indicating the areas to be worked on
2. *stop-and-go*—stopping the film, returning to the section to be discussed, and after comments have been made, continuing forward
3. *reel by reel*—stopping at the end of each reel for comments

Your editor will rely on you to take accurate notes during the running. Your responsibility for this will be lessened if you are required to operate the panel at the rear of the room, and if the editor and director choose to sit in the front of the room. If you have difficulty writing legibly during projection, try to rewrite the notes clearly for the editor before she begins making the director's changes.

Standard procedure is that during screenings the assistant *never* offers any suggestions or comments about the film or the editing and should only mention the amount of time that needs to be lost or added. Speak only when you are spoken to directly.

Even when you are directly questioned, be extremely diplomatic about the editing. You must always be protective of your editor and not *open up a can of beans* by some ill-considered remark.

The time between the last day of shooting and the first cut, or director's cut, screening should depend upon the amount of film the editor has been given. The more film the more time the editor should have, but television air and theatrical release dates rarely make this possible. In addition, these dates will restrict the overall editing time and other post-production work as well. You and your editor may have to work longer hours including weekends, and sometimes an additional editor and assistant have to be assigned to the project.

CHAPTER 13

THE FINAL CUT

Many directors may not have the time to see the changes your editor has made for them because they are already busy with their next assignment, probably on another series at a different company. Those who are available usually see their changes in the second cut, which is the first screening for the producer. With the impending air dates for weekly episodic television, having both the producer and director attend the second cut to discuss and agree upon changes will usually save valuable editing time.

Depending upon the pressure caused by air dates, film problems, and other factors, your second cut may be scheduled a day or two after the director's cut. Barring such unforeseen difficulties, the producer will normally see the picture at most two or three times before you will screen the *final cut* for him. All of these screenings occur over a period of about seven working days.

Your duties in assisting your editor before these screenings will be similar to those you performed before the first cut—measuring footage, checking for sync, finding and refiling trims, and any necessary splicing or repairing of film. If you did not have time to get music or sound effects for the first cut, try to do as much as you can for the second cut, concentrating on those areas where, according to your editor, it is really imperative.

FILING SCREENING TRIMS

Editors differ in the manner in which they prefer those trims that are the result of screening changes be filed. Some editors want the assistant to file each trim with its proper trim roll, but this requires disassembling splices when more than one cut is involved. The disadvantage of this is that each cut has to be retrieved from its individual trim roll and rejoined in the event that the editor wants to replace the trim back into the picture as originally cut.

The most satisfactory system is a combination of the above and a separation by scenes as was suggested in Chapter 12. If the trims consist of only one or two cuts, file each in its proper trim roll. All other trims should be left intact and filed within certain scene groups as *cut trims.* For example, assuming that your episode is slated through scene 80 you could have eight separations—scenes 1 through 10, 11 through 20, 21 through 30, and so forth. You do not want any cut trim roll file to be too cumbersome so the number of separations and how you separate them is your prerogative. Should one group become too large you can always divide it. If you can keep the trims in scene order within each group, you will be able to find a trim, should your editor need it, more quickly than if the trims are not in scene order. These trims are filed in identified boxes that are separated from the other trim boxes. Cut trims from subsequent screenings are simply added to the boxes.

A cut trim can sometimes consist of much more than several cuts. It can, in fact, be a section of a scene or an entire scene. It would then be considered a *lift*, in which case you would roll up picture and track as a separate entity, tab it with its scene identification and *measurement,* and file it in a separate lift box. The removal of this material should be indicated in the lined script, and you should add the measurement and running time. Should there be a need to add footage later, this record of the lifts can be most helpful to your editor.

MORE CUTS AND RECORDS

The screening record for the second cut (Figure 12-2) indicates that, in making the director's changes, your editor has lost over 2½ minutes and has been able to consolidate the picture on six reels instead of seven. The record also shows that both director and producer attended the screening along with several others, including the post-production supervisor who ordinarily would not have the time. However, in this case he might have wanted to attend in order to become more familiar with a new series and the work of a new editorial duo—you and your editor.

The procedures and your responsibilities for the second cut and further screenings are consistent with those for the director's cut. When you are screening for more than one individual, always indicate in your notes the person (director, producer and so forth) who has made a comment.

NETWORK INPUT

A representative from your series' network will attend one of your screenings. He will scrutinize your episode for any excessive or objectionable violence, sex, or dialogue. All networks employ such individuals who are referred to as *standards and practices persons* rather than as *censors.* Any changes they suggest will be made by your editor with the approval of the producer.

PREPARATION FOR NEGATIVE CUTTING

Although at this point your picture is not ready for negative cutting, it may save you valuable time later if, as you check sync on your picture before the various screenings, you simultaneously confirm that there are 16 frames to each foot as indicated by the key numbers within a *cut,* meaning here a section of a take that the editor has edited into the picture.

First you must understand that the negative cutter uses key numbers to match the appropriate negative to your work print exactly to the frame as the picture has been edited. Should your editor inadvertently *lose* one or more frames between two key numbers in a particular cut and it is not detected before going to the negative cutter, he might repeat the error when cutting the negative. The result would be a disturbing jump, or *jump cut,* in the completed product. In most cases the negative cutter would probably detect the error and notify the editor. The editor would then have to make some sort of correction determination.

How can an editor *lose* one or more frames? To illustrate this let's envision a hypothetical editing situation that has three cuts:

1. Anne says something to her friend, Larry
2. Silent cut of Larry reacting
3. Anne resumes talking (see Figure 13–1)

The *last* frame of the first cut of Anne (cut 1 above) is at G27X29325 + 08, and after the reaction cut of Larry, the *first* frame of the next cut of Anne (cut 3 above) is at G27X29333 + 01. Anne's trim, G27X29325 + 09 through 333 + 0 is filed in the

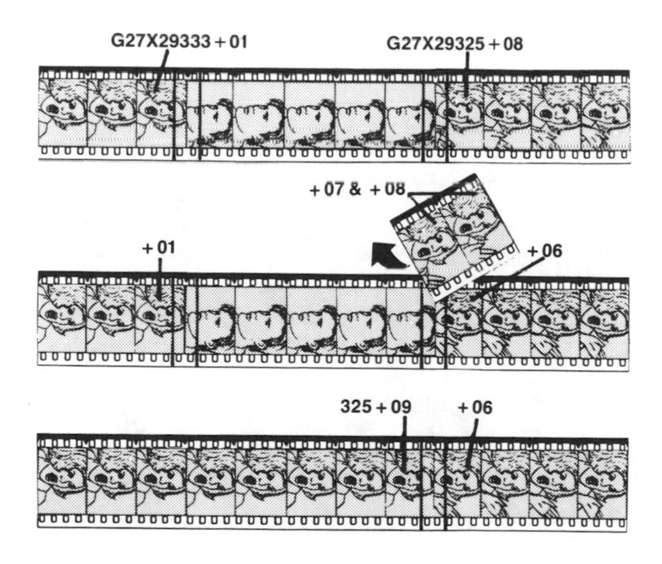

Figure 13–1. Losing two frames.

proper trim roll. After the first screening your editor decides the first cut of Anne (cut 1 above) is a bit long and takes off two frames of picture and track thus making the last frame G27X29325 + *06*. Your editor carefully marks the nearest code number with a grease pencil on the two frames (G27X29325 + 07/08), and places them in the short trims box or envelope.

A screening later the producer asks your editor to drop the cut of Larry reacting and stay on Anne through her dialogue. At the time the editor is making this change, you are both rushing to prepare for another screening. You find Anne's trim for your editor who removes the cut of Larry and replaces it with Anne's trim but forgets about the two frames previously removed. The first frame of the trim is G27X29325 + 09, which is now spliced to G27X29325 + 06. Frames 07 and 08 are missing thus creating a jump, or *jump cut,* in the film. If the editor had taken a few moments to count the frames between G27X29325 + 0 and G27X29326 + 0, he would have found only 14 frames and realized that the 2 frames were missing.

Such errors can be detected by the assistant in the same way if he checks wherever there are splices *within* a cut to confirm the 16 frames from key number to key number or code number to code number across the splice. A splice *within a cut* means a splice within a single angle. In the above example, the splice to be checked was in Anne's angle—from Anne to Anne. This would not apply to a splice from one angle to another such as from Anne to Larry. When you confirm the 16 frames, place a white grease pencil line across the splice on the black edge of the film to show that it has been checked.

In addition you and your editor must always be alert to the necessity of having to lose at least *one frame* whenever a take is intercut. This is because a *hot splicer* (see Figure 4–11) is used to join cuts in negative cutting. This splicer requires the loss of two sprockets from each side of the splice, thus negating a full frame from each scene.

In the situation example in Figure 13–2 there is a gunfight in which our hero shoots villain 1, and then shoots villain 2. The editor decides to intercut the hero shoot-

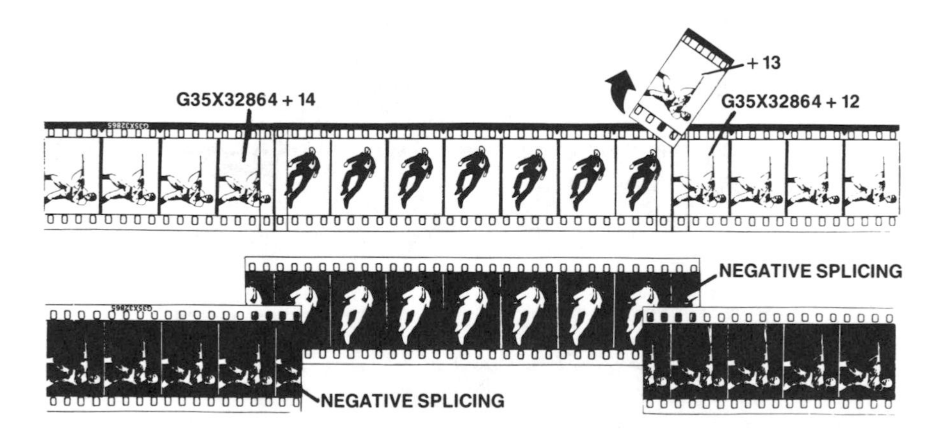

Figure 13–2. Obligatory one-frame loss.

ing so that we *see* the first villain falling, return as soon as possible to the hero firing his second shot, and finally see villain 2 falling.

The last frame of the first cut of the hero is at G35X32864 + 12. After villain 1 falls, the earliest frame the editor can use to return to the next cut of the hero is at G35X32864 + 14. Because the editor must lose at least one frame to allow for negative splicing, the thirteenth frame must go into the short trims box. The vigilant assistant will identify these types of cuts and make certain that the editor has *lost* the obligatory frame.

THE NETWORK FORMAT

The most obvious structural difference between television fare and theatrical films is that television stories are periodically interrupted by commercials, station identifications, and other material just as plays are interrupted by intermissions. Thus television movies, like plays, are divided into *acts* or *act breaks.* A half-hour episode has two acts; a one-hour show, four acts; and a two-hour movie usually, seven or eight acts. When a theatrical film is to be shown on television, an editor assigned to the syndication department will make similar act breaks at about 15-minute intervals depending on what is happening in the film. An act break should not interrupt a dialogue scene or action sequence in a manner that might disturb the viewers.

Some half-hour or one-hour series will have an extra act break when there is a *tag* after the last act. A tag is an epilogue that rounds out the preceding story and is usually about three minutes in length.

According to page one of the network format (see Figure 13–3) for your hypothetical one-hour series there are four acts but no tag. Sometime during the editing process when you have some time you should *build* in individual rolls the sections of material required by the format so that you can cut them into the picture for the final cut. This format material commonly referred to as *garbage* must be incorporated at the head and tail of the picture and along with the commercial banners at the act breaks. The reason this footage, consisting of mostly temporary leader, is not cut in earlier is that the director, producer, and editor do not want to sit through all that nonaction leader in all the screenings.

Several commercial banners, including the preceding and following black leaders, have been in the picture since before the first cut (see Chapter 11). These consist of those banners between acts I and II (9 through 11 in Figure 13–3), between acts II and III (13 through 15), and between acts III and IV (22 through 24).

Since you are beginning a new series, you will probably not have any actual picture with which to build these individual rolls. Such standard format material as the main title, bumpers, and logo will eventually be available to you so that you will include them instead of leader when you are building the garbage for future episodes. You will be working only with black and plain leader and commercial banners. The first commercial banner (5 through 7) is included in the first roll and there are two more commercial banners (27 and 30) that will have to be built into the last roll because they are not at act breaks and therefore were not previously inserted.

Page 1 of 2 -

FORMAT: ONE HOUR

	FORMAT MATERIAL	35MM FTG.	NETWORK INSERT	FTG.
1.	Black	2		
2.	TEASER	45		
3.	Black	2		
4.	MAIN TITLE	90		
5.	Black	3		
(6.)	Commercial Banner	1	Commercial #1	90
7.	Black	3		
8.	ACT I (Opening credits Supered)			
9.	Black	3		
(10.)	Commercial Banner	1	Commercial #2	90
11.	Black	3		
12.	ACT II			
13.	Black	3		
(14.)	Commercial Banner	1	Commercial #3	90
15.	Black	2		
16.	Bumper (Mx Over)	10+8		
*(17.)	Station Break Banner (SEE NOTE B)	1	Station Break	81
18.	Black	2		
19.	Bumper (Mx Over)	7+8		
20.	Black	3		
21.	ACT III			
22.	Black	3		
(23.)	Commercial Banner	1	Commercial #4	90
24.	Black	3		
25.	ACT IV			
26.	Black	3		
(27.)	Commercial Banner	1	Commercial #5	90
28.	Black	2		
(29.)	Trailer Banner (SEE NOTE A)	1	Trailer	90
(30.)	Commercial Banner	1	Commercial #6	90
31.	Black	2		
32.	END CREDITS	40+8		
33.	LOGO	5+8		
	TOTALS:	246		711

EACH ACT WILL OPEN AND CLOSE WITH A FADE

A

Your hypothetical format calls for the opening credits to be *supered* (superimposed) over the action footage of act I. Thus they are not part of the garbage or of any rolls to be built. See Chapter 15 regarding opening credits.

When you are building the format rolls, complete your picture material for a roll, then align one equal length of track leader for corresponding sync. Build footages *exactly* as indicated in the format with no extra frames.

The First Format Roll

Refer again to Figure 13–3. The first format roll consists of items 1 through 7. You begin with 2 feet of black leader followed by 45 feet of plain leader for the teaser,

Page 2 of 2

ONE HOUR

CUT BASIC SHOW TO:	4187	46 Min.	31-1/3 Sec.
Add Format Material	246		
DUB COMPLETE SHOW:	4433		
Lift Banners (-)	8	49 Min.	16 Sec.
Add Network Insert	711		
Total Air Time	5136	57 Min.	4 Sec.

The Editor will check to see that the Format Material is correct and the show is on footage before it is turned over to Music and Sound Effects. Any deviation from the Format must be reported to the Post Production office before the show is turned over.

Total screen time for the combined Executive Producer, Producer, Creator, Writer and Director credits must not exceed 10 seconds.

NOTE A: The Trailer Editor will provide a separate work print (action & track) to include Item #29. This film must be integrated into the work print of the show before it is turned over to Music and Sound Effects.

* NOTE B: Item #17: BEFORE any Format Material has been cut in, (measured from first frame of Reel #1), the total footage of ACTS I and II must fall between 1810 and 2710 feet.

AFTER all Format Material has been cut in, the Station Break Banner must fall between 1980 and 2080 feet as measured from the first frame in Reel #1.

B

Figure 13–3. Sample network one-hour format. (A) Page 1 and (B) page 2.

2 feet of black leader, 90 feet of plain leader for the main title, 3 feet of black leader, a one-foot commercial banner and 3 feet of black leader. Thus the first roll should measure a total of 146 feet and will be cut into reel 1 between the end of the academy leader and the first frame of the picture, the beginning of act I. Temporarily place the roll and corresponding track on one of the bench shelves separate from your other film.

The *teaser* in your format is thirty seconds of various sections of the episode. These cuts are carefully selected and orchestrated highlights, designed to entice the viewer to *stay tuned.* Who edits this teaser? If there is a *trailer editor* in your studio or company, he cuts the teaser. If not, then your editor will have the responsibility.

As your editor's confidence in you increases, he may let you cut some of the teasers, offering you an excellent opportunity to enhance and demonstrate your skills at editing and story telling. Materials used to cut a teaser are B&W reversals and dupe tracks of the picture. When the producer approves the teaser, either the trailer editor or you orders the necessary film (see Chapter 15).

The *main title* has the name of the series and introduces the stars, the *running,* or continuing, performers, and possibly the series' creator(s) and executive producer. Since this main title will appear on every episode and should establish the theme and style of the series in addition to introducing the performers, much attention is generally devoted to its creation. Material may consist of specific film shot prior to principal photography, cuts selected from the earlier episodes, sketches by an artist, animation, and titles created by a *title* person or by the title department of a lab. Once it has been approved by the producer, the same main title will be shown with each episode of the series.

Depending on the amount of available time, your editor or one of the other two editors on your series may be assigned to cut the main title. Otherwise another editor or title person will be assigned.

The Second Roll

Numbers 13 through 20 on your format is material required between acts II and III, halfway through a one-hour program. This is time allocated for local stations to identify themselves and is therefore called the *station break.* The commercial banner, 13 through 15, is already in the picture. Thus the second roll begins with 16, a 10½-foot *bumper* with MX (music). A bumper is usually a still picture that identifies the series such as the license plate on *L.A. Law,* the mansion on *Dynasty,* or the book on *All My Children.* A bumper also has either a musical background or a voiceover saying, for example, "(name of series) will return after the following announcement." The same bumper is reshown in all the episodes. These bumpers and tracks are often ordered for you and the other two assistants in sufficient quantity to last through the season.

Number 17 is for the *station break.* If you do not have a banner for this, just splice in a one-foot leader identifying it with your black marker. Numbers 18 through 20 includes a slightly shorter version of the previous bumper.

Your second roll, 16 through 20, should measure exactly 24 feet. Place it on the shelf next to roll one.

Roll Three

Since 22 to 24 between acts III and IV consist of a commercial banner that has been in the picture since the first cut, the third roll begins at the end of the picture with a commercial banner (26 through 28) followed by leader for the *trailer.* Similar to a teaser but twice as long, a trailer shows the highlights of the *next* episode of your series (". . . at this same time next week . . .").

Who will edit this trailer? Usually, a trailer editor does it. If not, the editor of the following episode will do it. Like teasers, trailers are cut with B&W reversals and dupe tracks of selected sections of the edited picture.

Number 30 calls for another commercial followed by *end credits.* As with opening credits, the producer must provide your editor with a list of various production, technical, and crew personnel for these credits. Usually the technical credits in the series' end credits, like main titles, do not vary from one episode to another but they still must be verified. Sometimes the end credits are supered over an established constant background repeated for each episode and are easily ordered. But often a producer will want the credits positioned over freeze-frame cuts of the preceding show thus reprising the leading characters at some exciting or humorous moment. A *freeze,* or *hold frame,* is a selected frame of picture that is optically reprinted for any desired length. Helping your editor select the exact frames will be additional but interesting work for you.

Roll 3 ends with the *logo,* 33, the symbol identifying your company as the producer of the series. Like the bumpers a number of these will be ordered for you and the other assistants. Place the roll on the bench shelf with the other two rolls.

You now have three individual rolls of format material to be inserted into the picture before the final cut. You will usually insert these rolls as you measure the picture and check it for sync. Most of this material is at this point in the form of leader, all carefully identified, to be replaced later with the actual film. When certain items such as the main title, the bumpers, and the logo are fixed and available, you may be able to incorporate them instead of using leader. Make sure you have corresponding leader for track for each roll and the music or narration tracks for the bumpers.

REBALANCING REELS

As the editor re-edits the film and makes the changes requested in the screenings, he takes footage out. At the director's cut your hypothetical picture was on seven reels and more than 18 minutes over the required footage. For the second cut, your editor had already removed, or lost, over 2½ minutes and had also *rebalanced* the reels and eliminated one reel. The picture was then on six reels.

Rebalancing means that the film is consolidated, and as film is added from one reel to another, it will mean changes at the end and beginning of some reels. These changes will naturally require new changeover cues as described in Chapter 12. The editor should rub out the existing changeover cues when the scene they were on is no longer at the end of the reel, and when he gives you the reel to splice, you must inscribe new ones. Sometimes the editor will forget to remove negated cues, and they will remain on a scene that may now be near the beginning or in the middle of a reel. You must clean them off to prevent an improper changeover in a screening. Always check the cues to verify that they are still valid or that new cues are required.

The editor must deal with another factor concerning the end of one reel and the beginning of another. He should try to begin each reel with a cut that has no extremely important action or dialogue in the first couple of feet. One reason for this is that, in case a projectionist errs in making the changeover, nothing vital will be missed. The other reason is a technical printing requirement for the finalizing of your picture regarding the combination of an optical track with the picture to form a composite print (see Figure 2–4). This single print containing both sound and picture runs in a projector that has the sound head in a lower area that is separate from the picture gate. If the optical track and picture were in sync *straight across,* the sound would be as late as the distance between the sound head and gate. Therefore, this distance must be compensated for by *advancing* the track before processing the composite print. Lab requirements are usually for an advancement, or *pull up,* of 20 frames for 35mm, 26 frames for 16mm. In editorial standard procedure is to allow for at least a 28-frame pull up, but it must be emphasized that you and your editor edit with sync straight across. The physical advancement of the sound is accomplished by the sound editors and the negative cutter in the final stages.

By selecting a particular cut or cuts to begin a reel, the editor is actually providing the necessary space for the future advancement of the sound. Your editor will try to begin a reel with a silent cut in this case meaning that it has no dialogue. Sound effects, even crowd background voices, may begin a reel. However, your editor will try to avoid *breaking* or *splitting* a reel, changing from one reel to another where he anticipates that there will be music. If the editor has been unable to avoid this, the music composer will try to hold a note or have a bridge, or break, in the music to cover this area.

As your editor gains confidence in you, he may leave the reels for you to rebalance.

FINAL SCREENINGS

The final cut has been scheduled and your picture, which is now on five reels, is either right on footage or within a few seconds under or over. Your editor knows where to add or eliminate those few seconds. Or the producer may take care of it with a few minor changes he may request so that when those changes are made and the picture is exactly on footage it is then considered *locked in* and you must take optical counts (see Chapter 14) and *turn over* the picture to M&E, music and sound effects (see Chapter 16).

Music Spotting

Usually as soon as possible after the final cut, the producer will have a *spotting session* with the music composer and the music editor. This screening is not usually attended by the series editor due to lack of time. Those who do attend determine where music should be placed in the picture to best help *tell the story.* As they identify these areas, they must also select, or *spot,* the exact dialogue or action where the music should begin and end. This is called *beginning* and *end cues.* They will also discuss the *kind* of music that should be written for certain scenes—a love theme for a romantic scene, a dramatic theme for an action sequence, and so forth.

There is a footage counter beneath the screen in the projection room. The projectionist will set the counter at 0000 at the 12-foot academy start before beginning each reel, and the music editor, in addition to noting descriptions of dialogue and action, will also record the exact footage in a reel for each music cue.

Sound Effects Spotting

The head of the sound effects department will attend the spotting session with the producer. He will represent all of the sound editors some of whom may also attend if they are able. Sometimes in order to save time the sound effects spotting will be combined with the final cut screening.

A secretary or a sound effects assistant may take notes and footages of specific effects requested by the producer. Most required sound effects will be self-evident and will be automatically cut in by the sound editor, whether or not it has been noted in the screening.

Network Screening

A final opportunity will be offered the network standards and practices section representative to object to anything he considers excessive violence, sex, or obscene language. The editor will generally attend this screening and report back to the producer. What changes should be effected will be the producer's decision.

THE ASSISTANT'S DUTIES

Although you would like to attend the additional screenings following the final cut to gain the additional experience, you may be unable to due to other responsibilities. During some of these screenings you may be fully occupied *bicycling* the reels between the projection room and your cutting room so you can order the opticals and send out some of the reels for duping. You might also be busy with dailies on your second episode.

Turning Over the Picture

When the spotting sessions have been concluded, you will be pressured by the music and sound editors for film to work on as soon as possible. The most common procedure is to order *two* B&W reversal prints of the picture from a facility specializing in reversals and *two* mag single-stripe, or *one-to-one,* transfers of the track from the sound department. In some companies additional tracks may be ordered for ADR and foley (see Chapter 16). The reversals are printed directly from your work picture (W/P), and transfers are made from your work track (W/T). It usually takes only a few hours to complete the work on all the reels of picture and sound.

When you get this material back, you should carefully identify each picture and track reel with the series name, production number, and reel number. Match the head pop at the 3-foot mark on the work track with those on the dupe (transfer) tracks, and aligning them in your synchronizer, back up to the 12-foot start of your work track. Mark corresponding 12-foot start marks on the dupe tracks. Use a piece of white tape to relabel the academy start marks on both the dupe tracks and reversal prints so that they might be more readily located. Mark the 12-foot start frame with a large *X* and write ''12-foot start'' beside it. Do not forget to use a red marker for sound and a black one for picture.

Now you are ready to turn the picture over to the music and sound editors. It is dispersed in the following manner:

1. You keep the work picture and a dupe track.
2. The music editor receives a dupe picture and dupe track.
3. The sound editors get a dupe picture, the original work track, and a copy of the code sheets. They also get the boxes of goodies and wild tracks.

The complete editing of your picture is now in the process of being finalized.

EXCEPTIONS

Feature Lengths

Features are not concerned with act breaks, commercials, or any kind of format material, at least not until they are shown on television. Nor is there any specific restriction in length. Some features run about 100 minutes; others are as much as 3 hours long. Producers, always observant of post-theatrical revenue, try to bring their pictures in at a length that can be conveniently adjusted later to the requirements of television.

Opening Titles and Credits

Instead of opening credits being a separate section as on your episode, in a feature, TV movie, or miniseries they are combined with the main title. Some may be done with a great deal of creative skill as for example the animated main title of the *Pink Panther* films; others may be done quite simply with titles and credits fading in and out on a black background.

Editing Time

Generally whether the film is for the theatre or television, the more extensive the project, the more time is required for editing. Episodic television is relatively quick—a few screenings and the picture is locked up. Most theatrical films and long-form television after shooting are edited for two months to as much as a year with constant screenings throughout.

The amount of time spent in post-production depends partly on whether the director remains in control of the editing throughout or if the director must contractually complete his perspective of the film in a short period such as one or two months following shooting at which time the producer assumes control. On these longer projects the composer usually attends earlier screenings and has more time to create some of the themes that will be needed.

Feature Ratings

While features have no standards and practices network representative to screen for until they are adapted for television, they must be viewed by the Motion Picture Association of America (MPAA) to establish a boxoffice rating. Should a feature receive a rating that the producer feels is undesirable for the product, he must negotiate with MPAA to determine the editing changes that are necessary in order to receive a better rating.

The assistant must keep informed about these screenings, have the film at the proper projection room on time, and take accurate, legible notes when it is necessary.

Miscellaneous Screenings

Because you are working on an episodic series, you and your editor constantly have another episode to work on immediately after the completion of one episode and are usually unable to attend screenings other than those for the producer and director. This situation does not exist on TV movies or features unless of course an editor's next assignment begins on the heels of a previous show. Thus the editor and assistant are usually able, and expected, to attend all screenings of their picture including spotting sessions and music scoring. This gives the editor an opportunity to offer some input and the assistant gains additional knowledge about these areas of post-editing.

Turnover Material

The number of reversal prints and dupe tracks ordered for music and effects varies somewhat from company to company. Although the cost of reversals and transfers are not considered expensive items in post-production, there are some companies that feel they should save some cost by ordering only one reversal and transfer of the picture, forcing M&E to share the material by bicycling the reels back and forth between them. Obviously this will make their work more difficult, create longer work hours, and possibly delay completion.

On the other hand some companies believe it is worth the extra expense to order additional dupes for the *ADR* and *foley* editors (see Chapter 16). In addition, it is now common practice to order a video cassette of the picture for the music composer. Before you order anything, ascertain your company's policy.

CHAPTER 14

OPTICALS

Opticals are visual effects that serve as transitions in time, place, or action or that heighten a dramatic or comedic scene and correct or improve direction or performance. Opticals include such effects as *fades, dissolves, wipes, flips, superimposures, freeze* or *hold frames, flopovers, tail-to-head reverses, blowups, move-ins, irises, repositions,* and *skip frames.* Opticals are created in a lab's optical department or in a specialized optical facility using production film already shot and an optical printer.

Opticals are ordered by the assistant and they are usually accompanied by *count sheets* (see Figure 14–1). Since the assistant often uses these same count sheets to order other material such as straight dupes, established main titles, and other titles or credits, these items are usually included in any reference to *opticals.*

As was discussed in Chapter 12, certain opticals may be ordered early in the editing but most opticals are best ordered after the final cut when the show is locked in and it can be assumed that there will be no further editing changes that would affect the opticals. It is vital that opticals be ordered as soon as it is possible to do so. However, at this particular time there may be spotting sessions after which you need to get B&W reversals and dupe tracks made for music and sound effects. You may also be involved with dailies on a new episode. You have to accomplish all this and also order opticals.

You can remove some of the pressure by writing up your count sheets *before* the final cut, especially when your show is close to the required footage. Hold the orders until after the final cut so that you can make any corrections made necessary by final editing changes. Your editor can help by advising you of a change being made that will affect an optical. Before you send any reel for duping make certain that it is in sync and that your footage measurement and optical counts are accurate.

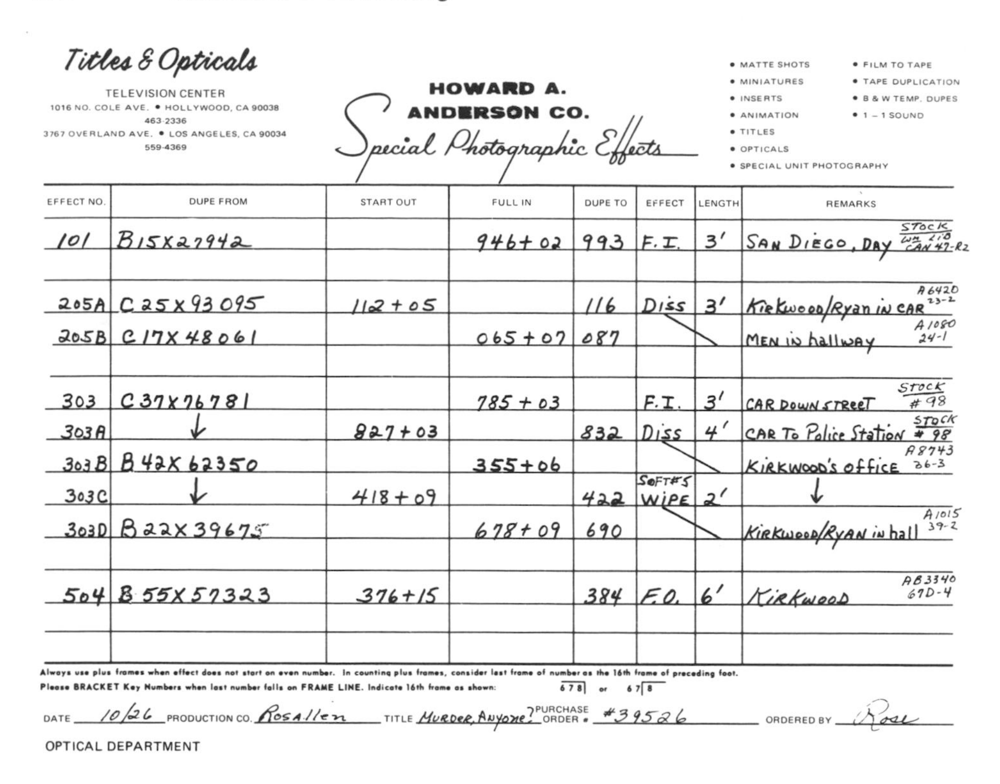

Titles & Opticals

TELEVISION CENTER
1016 NO. COLE AVE. • HOLLYWOOD, CA 90038
463-2336
3767 OVERLAND AVE. • LOS ANGELES, CA 90034
559-4369

HOWARD A. ANDERSON CO.
Special Photographic Effects

• MATTE SHOTS • MINIATURES • INSERTS • ANIMATION • TITLES • OPTICALS • SPECIAL UNIT PHOTOGRAPHY
• FILM TO TAPE • TAPE DUPLICATION • B & W TEMP. DUPES • 1 – 1 SOUND

EFFECT NO.	DUPE FROM	START OUT	FULL IN	DUPE TO	EFFECT	LENGTH	REMARKS
101	B15X27942		946+02	993	F.I.	3′	SAN DIEGO, DAY STOCK WA 210 CAN 47-R2
205A	C25X93095	112+05		116	DISS	3′	Kirkwood/Ryan in CAR A6420 23-2
205B	C17X48061		065+07	087			MEN in hallway A1080 24-1
303	C37X76781		785+03		F.I.	3′	CAR DOWN STREET STOCK #98
303A	↓	827+03		832	DISS	4′	CAR To Police Station STOCK #98
303B	B42X62350		355+06				KIRKWOOD'S office A8743 36-3
303C	↓	418+09		422	SOFT #5 WIPE	2′	↓
303D	B22X39675		678+09	690			Kirkwood/Ryan in hall A1015 39-2
504	B55X57323	376+15		384	F.O.	6′	Kirkwood AB3340 67D-4

Always use plus frames when effect does not start on even number. In counting plus frames, consider last frame of number as the 16th frame of preceding foot.
Please BRACKET Key Numbers when last number falls on FRAME LINE. Indicate 16th frame as shown: 678] or 67[8

DATE 10/26 PRODUCTION CO. Rosallen TITLE Murder, Anyone? PURCHASE ORDER # #39526 ORDERED BY Rose

OPTICAL DEPARTMENT

Figure 14–1. Optical count sheet.

In order to understand what you must order for opticals you must understand the requirements of negative cutting. The negative cutter matches the original production negative to the work print by *key numbers* and assembles the negative exactly as the editor has edited the work print. Since making an optical involves the original negatives, opticals must be ordered and made before the negatives are cut. If for example the editor wants to repeat a particular piece of action, a reprint can readily be ordered from the original negative. Eventually, however, the negative cutter will have to be provided with a *duplicate* negative of the repeated section. Once a section of a particular negative has been cut and spliced into the negative picture that piece of *original* negative cannot be used a second time.

All opticals must be given to the negative cutter in the form of *dupe* negatives. The assistant receives dupe prints of the opticals, and these are used to replace the original prints in the picture. Then the negative cutter matches the dupe negatives to the dupe prints.

THE DUPING PROCESS

Figure 14–2 demonstrates how the dupe negative and dupe print can be achieved in both black and white and color film. Whether it be B&W or color film, the daily prints

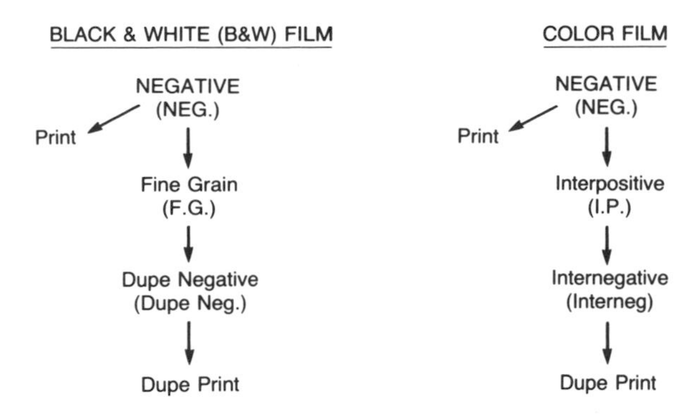

Figure 14–2. Duping negatives.

are from the original negatives. In B&W film a *fine grain* (F.G.), a specially processed print from the negative, is made. From that a *dupe negative,* is made and then a *dupe print.* The same process is performed in color film with only a change in terminology, from original color negative to *interpositive* (*I.P.*) to *internegative* (*interneg*) to *dupe color print.* The term *dupe negative* is generally used to mean either the B&W dupe negative or the color internegative.

Since a negative is a precious commodity, the less it is handled the safer it remains. Once an interpositive is made it protects the negative because you can order numerous internegs from the I.P. without involving the original negative. A fine grain similarly protects a B&W negative.

DUPE KEY NUMBERS

During thc duping process the key numbers on your dupe prints will change from *original* key numbers to *dupe* key numbers. The original Eastman number G37X48219 for example will remain through the I.P. stage but then will change to a dupe number on the interneg. The dupe key number will appear on the dupe print. Until recently an example of a dupe number was 30X72305, beginning with a numeral rather than a letter. You may still come across dupe stock with such key numbers, but the newer dupe numbers resemble original numbers; they begin with an *A* or a *D*, for example A30X72300 or D37X48200.

When original Fuji film negative is used, the daily print is usually on Fuji stock, but it is duped on Eastman stock because Fuji does not produce the required stock for duping. Therefore, do not be surprised when you find *both* Fuji and Eastman identifications on your dupe prints—original Fuji key numbers with Eastman dupe key numbers.

In addition the original Eastman key number on Eastman stock may sometimes be retained throughout the process so that you might have both dupe and original key numbers on an Eastman dupe print. This can be helpful to the assistant when she is matching and cutting in the dupes although it may momentarily confuse the negative cutter.

The negative cutter uses whatever key numbers (either original or dupe) you have on the work print to match and build her negative reels in precise duplication of your reels.

STRAIGHT DUPES

A *straight dupe* is a section of film with no optical effect whatsoever that has been duped. Why then has it been duped? One reason is to save the original negative from being cut so that the material may be used in the future. Examples are a piece of stock, like a car runby, an establishing shot of a certain location, or a war scene.

A second reason for a straight dupe is the repetition of a cut, a shot the editor wishes to use later in the picture that she had already used in an earlier reel.

For example in a hypothetical situation there is a single cut in reel 2 of Kathleen sleeping in bed, awakening, and exiting the bedroom. The editor wants to use only the head section, a short cut of her sleeping, in reel 5. It is no problem for you to order a reprint from the original negative for the cut in reel 5, but this is only a temporary solution. Now you have two sections of the same scene with the same key numbers in two different reels. Once the negative cutter has matched and cut the negative for reel 2, if the reprint is still in reel 5 she has no matching negative. Before the negative cutter can complete reel 5, you must dupe the repeated section of film by ordering an I.P., interneg, and dupe print. The negative cutter receives the interneg to use in matching to your dupe print.

In this example there might be another possibility, one in which an optical would not have to be ordered. If this scene had an outtake, or *B* negative that could be ordered and was usable, it would replace the reprint in reel 5, and the negative cutter would match it with the original negative.

Trailers and teasers, which were discussed in Chapter 13, consist of various cuts from a film. The original negatives are used to match those scenes in the film so the duplication of them in trailers and teasers will require duping. Since optical effects are not usually involved, these trailers and teasers are also straight dupes. The main title and logo for a series usually involve some effects in the initial duping process but once they are set, they are considered straight dupes when they are reordered for each episode.

Straight dupes are included with your count sheet orders for other opticals.

EDITOR'S OPTICAL MARKS

I suggest that all original prints that are designated for stock be red lined before being given to the editor. A reprint should also be red lined if it is a temporary print for a repeated cut, i.e., it will eventually be replaced by a dupe print. When an editor asks you to order a reprint, determine how she is going to use it. If it is to replace damaged film, it should not be red lined unless the damaged film has been red lined.

If you neglect to red line a take when the editor uses part of the take and realizes your error, she will probably draw a white grease-pencil line down the center of the cuts she has used. When you splice or check sync and footage, you should rub the white line off and red line the cut. Also, red line the trims of the shot in case the editor uses any of the trims later.

Identifying Opticals

As your editor is editing from first to final cut, she will inscribe other markings for a variety of optical effects. You must leave these markings on the film for the benefit of the director or producer in a screening when they will be told what kind of effect is being proposed and for your benefit when you order the opticals by *taking counts.*

Opticals such as fades, dissolves, wipes, and flips are easily distinguished by the manner in which the editor marks the film. Some editors carefully mark these opticals exactly to the length of the effect desired whereas others mark general lengths and expect the assistant to know what was intended and to mark it precisely. Figure 14–3 shows how several of the most common opticals may be marked on the film by an editor. Read the film as you would in your synchronizer from right to left and head to tail.

For example, if an editor has casually lined a fade for 2 feet + 10, the assistant will need to correct it to exactly 3 feet + 0 because such opticals are normally in round figures. If the editor had wanted a 2½-foot fade or any other deviation from the norm she should have written the footage on the film. It is always a good policy for an editor to indicate unusual footages or effects for a new assistant or for one unfamiliar with her practices so that the assistant can properly complete the footage.

Actually, it is practical not to take the time to correct the footages until you are ready to order opticals because changes requested during the various screenings might modify or even eliminate effects. At the appropriate time after she has verified the exact footage on each optical, the assistant *closes,* or *blocks, in* the editor's lines (see Figure 14–4) at the last frame of the fade in (b), at the 1st frame of the fade out (a) and at the 1st frame (a) and last frame (c) of the dissolve and wipe. This is done simply by extending the grease-pencil lines along the picture frame line at those points from edge to edge.

To distinguish more easily between a dissolve and wipe instead of a top to bottom diagonal line from (a) to (c) for a dissolve some editors may indicate the wipe by lining

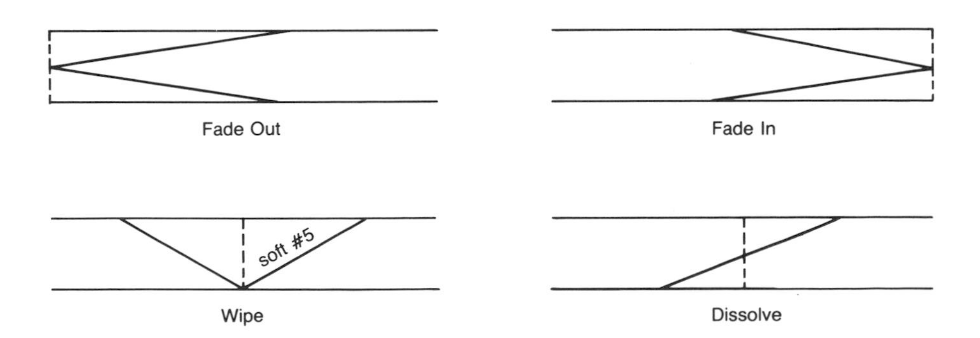

Figure 14–3. Editor's optical marks.

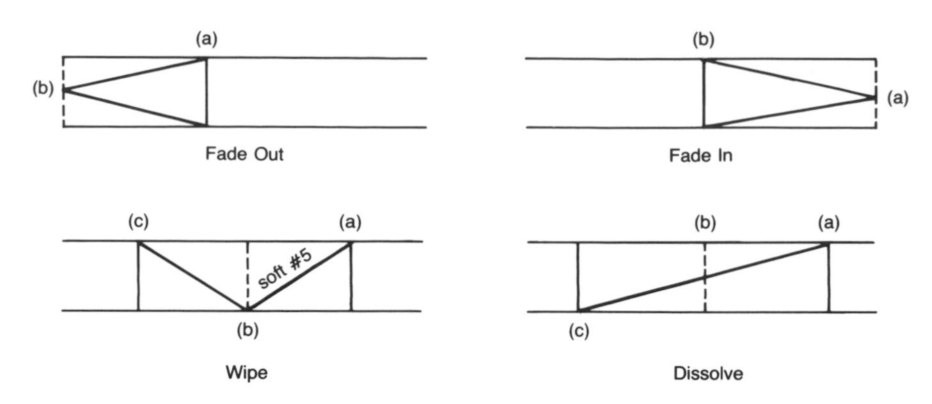

Figure 14–4. Closing in optical footages.

it from top (a) to top (c) with the bottom of the center splice (b) as the apex of an inverted triangle as shown in figure 14–4. There are a number of variations for most effects. An assistant must learn her editor's markings.

MAJOR OPTICALS

A *fade in* (F.I.) is an effect in which the scene starting in black gradually appears. The point, or frame, at which it is fully and completely discernible—(b) in Figure 14–4—is referred to as *full in* (F.I.). The use of fade ins is commonly restricted to the beginning of a picture and in television to the beginning of acts.

A *fade out* (F.O.), the reverse of a fade in, is when a scene gradually disappears, ending in black. The frame that begins the effect (a) is the *start out* (S.O.). A fade out is usually used at the end of television acts and at the conclusion of a film.

Dissolves are the most frequently used effect for establishing transitions in time, place, and action. A dissolve consists of both fades superimposed. As the outgoing, *A,* scene fades out, the incoming, *B,* scene fades in and overlaps the outgoing scene. The first frame (a) of the dissolve is the start out and the last frame (c) is the full in. It is important to understand that the *B* side must begin fading in at the start out of the *A* side and that the fade out of the *A* scene ends at the full in of the *B* scene. When you receive the dupe print containing the dissolve, the splice (b) between the two scenes will have become the center of the effect.

Most dissolves, particularly in episodic television, are two or three feet in length. But they can be any length an editor wishes consistent with the style, mood, and pacing at that moment in the picture. A dissolve of less than a foot is sometimes used to *smooth out* a disturbing cut that is then called a *soft cut.*

A *wipe* is another effect that serves a purpose similar to that of a dissolve. Instead of overlapping scenes a wipe uses the incoming scene by means of a *hard-edged* or *soft-edged line* to literally *wipe off* the *A* scene as the *B* scene becomes solely and completely visible.

The optical facility will provide your editor with a chart or list of various wipes that may be ordered by number. Most optical houses can do over 100 different kinds of wipes. Your editor must advise you which (number) wipe she wants and whether they should be hard- or soft-edged. A wipe, like a dissolve, begins with a start out and is completed at the full in.

A *flip* is another transitional effect that begins with a start out and ends with a full in. While dissolves and wipes are usually more than a foot long, from S.O. to F.I., a flip is usually only eight frames. It features a widening black stripe either on both sides or on both tops of the picture frame that squeezes out the *A* scene and conversely brings in the *B* scene thus creating an extremely fast turnover, or revolving, effect. It can be made left to right, right to left, top to bottom, bottom to top, or diagonal in direction. Again, your editor must let you know exactly which kind of flip she wants. A flip is usually used for comedic effect or to speed up transitions especially in action sequences.

TAKING COUNTS

Taking counts is the ordering of opticals by the use of key numbers. If your editor believes there will be no further changes in the area of the opticals, she may authorize you to take counts on them before the final cut. Taking counts at this point will relieve some of the pressure you will have after the screening when you have to help your editor with implementing any final cut changes and getting the picture on footage, *turning over* the picture to music and sound effects (as was discussed in Chapter 13), and possibly syncing and breaking down dailies on the next episode.

You can adjust your counts accordingly should there be some optical change as a result of the final screening.

Count Sheets

Count sheets are the order forms by which assistants order their opticals. They vary from one facility to another but they require basically the same information:

1. What kind of effect is it?
2. What is its length and where should it begin and end?
3. Where should the film before and after the effect begin and end?

On some count sheets the beginning and end of the entire optical is referred to as *start printing* and *stop printing*. On your form (see Figure 14–1), the start printing is *dupe from* and the stop printing is *dupe to*. This information is rounded out to the nearest full key number and does not contain plus frames.

Extra film footage, called *handles,* is normally added to the beginning and end of each optical. Usually one or two extra feet are added, but this is dependent upon your company's policy. On your hypothetical count sheet *one-foot* handles are ordered. This extra footage is for protection and insures that enough footage is being ordered. Handles also provide a margin by which the editor might slightly alter the optical should she feel it necessary.

The *full* original key number of each shot used in an optical must appear *once* on your order. The amount of space given on the form you are using will indicate where the optical department wishes you to write the full original key number. On Figure 14–1 it is the *dupe from* column. Other forms may give you sufficient space under the *start out* or *full in* columns. In the other columns on the count sheet you need only list the last three numerals of the key number plus frames when that is necessary. Plus frames should *only* be used for the *start out* and *full in* counts.

Counting Frames

Most mistakes on count sheets are made in counting the frames from the nearest key number to the *full in* or *start out.* In either case, because frame identification is always in *plus frames* and *never* in minus frames, the nearest key number to the full in or start out would be the one just before (or to the right going from head to tail). You will always count frames right to left.

With your film emulsion side up you have to read Eastman key numbers in reverse order right to left and backwards. If, for example, the Eastman key number is G42X83967, regardless of how you have to read it, the last three numbers will always be 967. The last number is *7,* and wherever that last number is, this is the *0* frame of the picture and the next frame reading right to left is the first frame, the next frame is the second frame, and so on.

A Fuji key number, unlike an Eastman number, is imprinted left to right. In Figure 14–5 the last three numbers are 871 and *1* is the 0 frame. You still count the frames from right to left. In the figure hypothesize that the *X*s are the start out or full in of an optical. What are the *counts* to and including that *X?*

For both Fuji and Eastman you count right to left. In Figure 14–5 the correct Fuji count is 871 + 09 and the Eastman count is 967 + 06.

What would the counts be if in both cases the last numbers were on the frame lines or so close to a frame line that you could not be sure which frame should be the 0 frame? In Figure 14–6, the last Fuji number, *1,* and the last Eastman number, *7,* are on, or so near, the frame lines that the 0 frame could be either the left or right frames. It is your decision which frame is the 0 frame and you must indicate it on your count

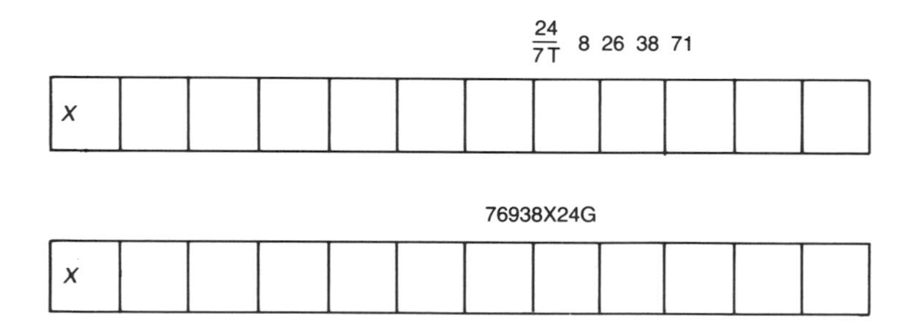

Figure 14–5. Counting frames.

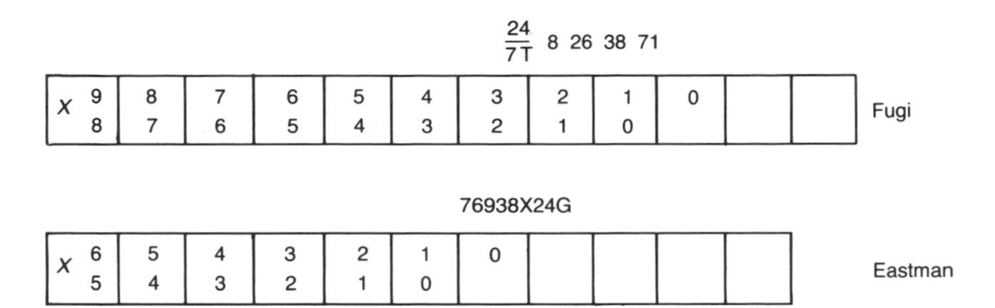

Figure 14–6. Frame-line counts.

Fuji: 871 + 08 or 87 1 + 09

Eastman: 967 + 06 or 96 7 + 05

Figure 14–7. Frame-line counts solutions.

sheets so the optical company will clearly understand. You do this by bracketing the last key number either with the rest of the key numbers or separated from them (see Figure 14–7).

Effect Numbers

In the first left-hand column of your count sheet you number the effect for purposes of identification. A common system is to let the number you assign to an optical identify where that optical is in your picture so that you can easily locate it when you receive the dupe print (the completed optical). You accomplish this by having the first number represent the reel number and the next one indicate the sequential optical in that reel. Thus 101 on the hypothetical count sheet would be the first optical in reel 1, 205 the fifth optical in reel 2, 303 the third optical in reel 3, and 504 the fourth optical in reel 5. (Obviously the counts shown in Figure 14–1 are isolated examples as they omit other opticals.)

Remarks

In the extreme right-hand column under *Remarks* give a brief description of the scene. Also note the code number so you can identify the scene and take number after you have completed the counts. Should a question arise concerning an optical, this information may be helpful to the negative cutter, the optical house, and you.

Continuous Counts

Optical 303 as it has been diagrammed in Figure 14–8 contains a series of continuous opticals that will become one dupe print when you receive it. The first optical, a 3-foot

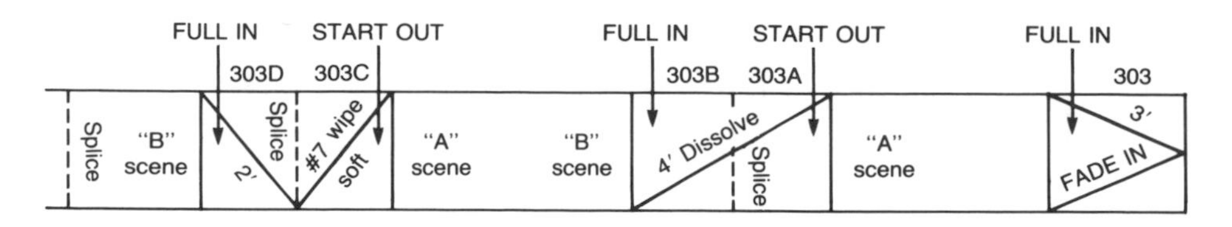

Figure 14–8. Optical 303 diagrammed.

fade in—unlike a dissolve, wipe, or so on—is a single entity; a fade in involves only one cut and is numbered 303. However, at the end of the cut it becomes an outgoing *A* scene, 303A, dissolving to an incoming *B* scene number 303B. Note that as 303B continues on to wipe to another scene, it changes from being a *B*, or incoming, scene and becomes an *A*, or outgoing, scene, *303C.* The *B*, or incoming side of the wipe, *303D,* ends the optical because it straight cuts (no optical) to the next scene.

A straight cut will end an optical. For example, if the first scene, 303—the fade in of the car runby to the police station—had not dissolved but had straight cut to Kirkwood's office, optical 303 would have ended at the cut. The wipe from the office to the hallway scene would then begin a new optical, 304A.

Outside/Inside Opticals

Dependent upon your company's policy and the type of project you are working on, either *outside* or *inside opticals* may be ordered. An outside optical is a dupe of the entire cut, which obviously includes the effect itself. On the count sheet (see Figure 14–1) *outside* opticals are ordered. For effect 504 over 60 feet (from 323 to 384) is ordered for only a 6-foot fade out; for 205, 21 feet for the outgoing scene and 26 feet for the incoming scene. This totals 47 feet for a 3-foot dissolve.

An *inside optical* is only sufficient footage for the optical plus the handles. If you were to order inside opticals instead of outside opticals, which counts would change on your count sheet in Figure 14–1? Do not forget the one-foot handles. Obviously, only your *dupe to* and *dupe from* counts would be affected by changing to inside opticals. Before you read further, can you figure out the changes?

Now look at the revised count sheet in Figure 14–9, which indicates ordering of *inside* opticals. Did you make the correct changes? By comparing these new counts with those in Figure 14–1 which indicates the ordering of *outside* opticals, you can see the amount of footage that need not be ordered. By ordering inside opticals for just these four effects, you would have ordered 245 feet less in Figure 14–9 than in Figure 14–1. In effect 101 you dupe to 947 instead of 993, a saving of 46 feet. You save a total of 36 feet in effect 205 and 52 feet in 504. The most drastic change is in effect 303. Besides saving a total of 111 feet, instead of one continuous optical you now have three individual effects.

Thus the reason for ordering inside opticals is purely economic; by eliminating footage you can save film and lab costs. In some companies the post-production supervisor may authorize the assistants to order inside opticals only when a certain amount

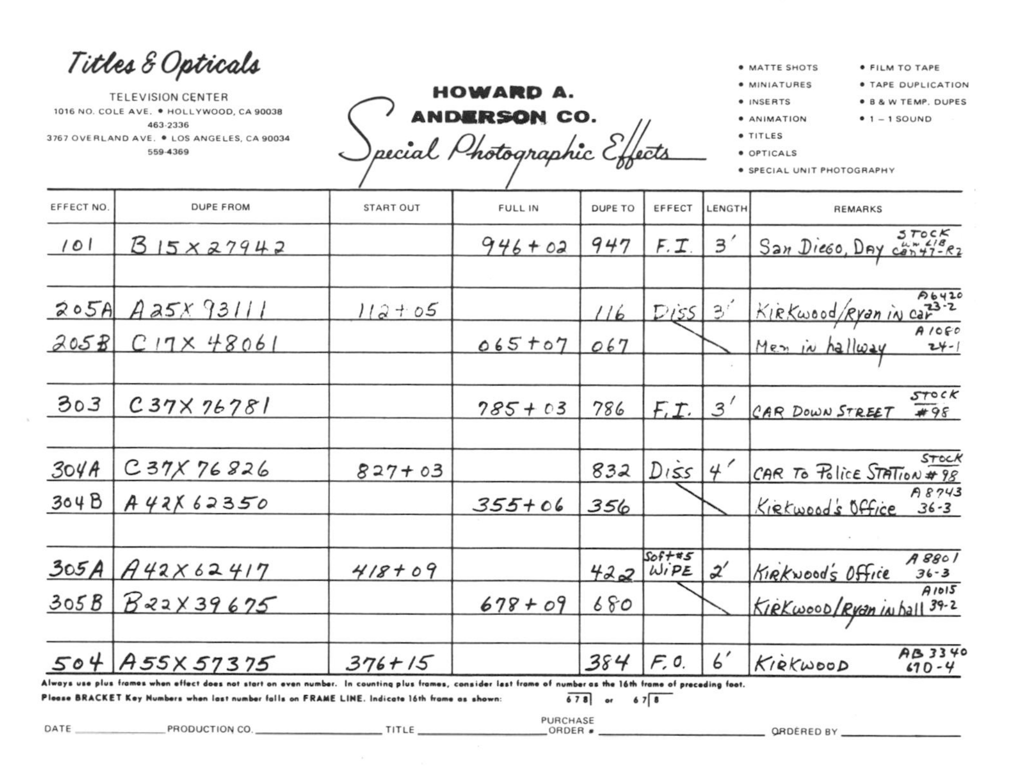
Titles & Opticals

TELEVISION CENTER
1016 NO. COLE AVE. • HOLLYWOOD, CA 90038
463-2336
3767 OVERLAND AVE. • LOS ANGELES, CA 90034
559-4369

HOWARD A. ANDERSON CO.
Special Photographic Effects

• MATTE SHOTS • MINIATURES • INSERTS • ANIMATION • TITLES • OPTICALS • SPECIAL UNIT PHOTOGRAPHY
• FILM TO TAPE • TAPE DUPLICATION • B & W TEMP. DUPES • 1 – 1 SOUND

EFFECT NO.	DUPE FROM	START OUT	FULL IN	DUPE TO	EFFECT	LENGTH	REMARKS
101	B15X27942		946 + 02	947	F.I.	3′	San Diego, Day STOCK car 47-R2
205A	A25X93111	112 + 05		116	DISS	3′	Kirkwood/Ryan in car A6420 23-2
205B	C17X48061		065 + 07	067			Men in hallway A1060 24-1
303	C37X76781		785 + 03	786	F.I.	3′	CAR DOWN STREET STOCK #98
304A	C37X76826	827 + 03		832	DISS	4′	CAR TO Police STATION STOCK #98
304B	A42X62350		355 + 06	356			Kirkwood's Office A8743 36-3
305A	A42X62417	418 + 09		422	Soft #5 WIPE	2′	Kirkwood's Office A8801 36-3
305B	B22X39675		678 + 09	680			Kirkwood/Ryan in hall A1015 39-2
504	A55X57375	376 + 15		384	F.O.	6′	KIRKWOOD AB3340 610-4

Always use plus frames when effect does not start on even number. In counting plus frames, consider last frame of number as the 16th frame of preceding foot.
Please BRACKET Key Numbers when last number falls on FRAME LINE. Indicate 16th frame as shown: 678] or 67[8

DATE ______ PRODUCTION CO. ______ TITLE ______ PURCHASE ORDER # ______ ORDERED BY ______

Figure 14–9. Revised counts for inside opticals.

of footage is saved on a single optical, i.e., more than 35 feet or 50 feet. Otherwise they must order outside opticals.

Why should full length opticals ever be ordered when it is considerably less expensive to order inside opticals? When you receive the dupe prints containing only the effects, you will have to cut them into your work print thereby replacing only the film with the optical markings. When the negative cutter matches these areas, there will then be an internegative spliced to an original negative *within the same cut* often resulting in a disturbing change in density or color. This discourages most directors and producers from approving the use of the shorter opticals. Inside opticals are often used in episodic series but rarely on longer projects unless the project has a limited budget.

Inside/Outside Counts

While inside/outside opticals concern where you start and stop printing the complete optical, an *inside* or *outside count* involves *only* the *full in* or *start out* counts.

The *start out* and *full in* counts on both the hypothetical count sheets are *inside* counts because you have designated them as the beginning or end frame *within* or *inside* the effects. In effect 101 the full in, 946 + 02, is the last frame within the 3-foot fade in and is therefore an inside count. In effect 205 the start out is the first frame, and the full in, the last frame of the dissolve. They are, therefore inside counts.

An *outside count* is counting to the frame *preceding* the first frame of an optical for the start out and to the frame *following* the last frame of an optical for the full in. If outside counts were being ordered on the effects, the full in for 101 would be 946 + 03. In effect 205 the start out would be 112 + 04 and the full in, 065 + 08.

Even though outside counts are rarely used, always find out your company's policy regarding these counts and make certain the optical company understands that policy.

Do not confuse *inside/outside counts,* which concern the full in and start out counts, with *inside/outside opticals* which pertain to the dupe from, or start printing, and dupe to, or stop printing, counts.

OPTICAL EXERCISES

Figures 14–5 and 14–6 are simulations of how you would actually count the frames for your full in or start out counts if you had the film in a synchronizer in front of you. Without film or synchronizer, with only paper, pencil, and a sample of the count sheet columns as shown in Figure 14–1, try to do the counts given the following information:

> For the fourth optical in reel 4, make a three-foot F.O. for a cut in which the first frame is H25X37215 + 10 and the last frame, 243 + 08; use an *inside count* for this *outside optical* with one-foot handles.

First study the diagram in Figure 14–10 so that you fully understand the information given you for this optical. You are looking at the imaginary film right to left, or head to tail. Imagine on the right there is another scene ending at the splice where the cut begins at H25X37215 + 10 and on the far left the last frame is 243 + 08, which is where the scene fades out.

Make a copy of the count sheet (Figure 14–1) by duplicating the columns on a sheet of paper. You are now ready to take counts on the above. Begin with the effect number. The first digit is the reel number, the other two digits indicate the sequential number of the optical in the reel.

The next column is *dupe from.* This column indicates where the optical house is to start printing your *outside* optical. Since you know the first frame of the cut, back up or subtract one foot for the handle or extra footage, and round off to the *nearest* key number. Remember to give the *full* key number in this column.

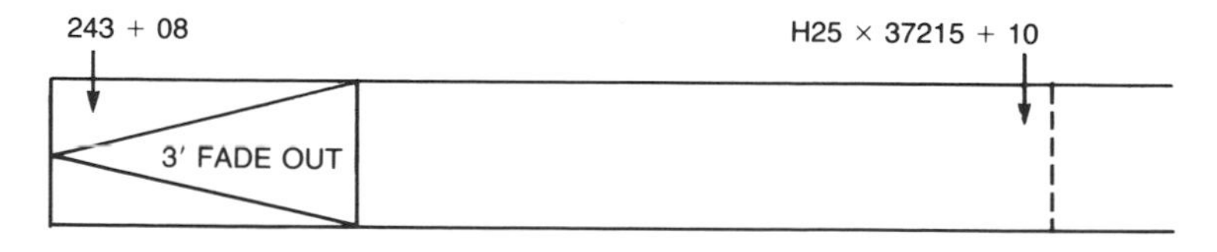

Figure 14–10. Fade out exercise diagram.

What kind of optical is it? Go to the right hand columns and under *effect* write "FO" for fade out and under *length* write three feet. Does a F.O. have a start out or fill in? Do you remember that a fade in has a full in and a fade out has a start out?

If you had the film in your synchronizer, you would roll down to the last key number before the S.O. of the F.O. and write down the last three key numbers in the start out column. Then you would set the framer on your sync machine at *0* frame at the *last digit* of the key number and roll down to the S.O. The framer would give you the plus frames to be added to the last three key numbers in the S.O. column.

You do not have the film and equipment to do this, but try to arrive mathematically at your start out *inside count,* which you had to measure carefully and *close in* at exactly three feet + 0. You know that the effect is three feet long and that the last frame is at 243 + 08.

The *dupe to* column is easy to fill in. Just add one foot for the handle to the last frame, going to the nearest full key number. Did you remember that you only use plus frames for the S.O. and F.I. counts?

The remaining column, *remarks,* gives you an opportunity to use your imagination. Be creative and hypothesize a scene that would be fading out, for example, at the end of an act in your episode. Do not forget to include code and scene/take numbers.

Now compare your counts with effect 404 in Figure 14–13. Do all your counts correspond? First you were told it was the fourth optical in reel 4. Did you identify it as effect 404? You were then given the first frame of the cut as H25X37215 + 10. Subtracting one foot for a handle would come to 214 + 10 so the nearest full key number for the *dupe from* column would be H25X37215. (Ten frames is close enough to a foot.) The last frame of the cut is at 243 + 08. When you add one foot for a handle, you arrive at 244 + 08. The nearest full key number is at 244 but that would give you only an 8-frame handle so you should have written *245* in the *dupe to* column.

Were all your counts correct except for the S.O. count? Did you have plus eight frames instead of plus nine frames? This is understandable. The last frame is 243 + 08, and when you subtract three feet for the effect, H25X37240 + 08 would appear to be the logical count. But it would be wrong. That + 08 frame that you begin counting back from is counted as one of the 16 frames comprising one foot. Each imaginary frame for the first foot back is listed below. Count back from the last *frame,* which is 08 but count that frame as the first frame.

08 07 06 05 04 03 02 01 00 15 14 13 12 11 10 09

As you can see, counting back one foot your sixteenth frame would end on 242 + *09,* not + 08. Therefore whether you are subtracting one foot, three feet, or ten feet from a plus eight frame, you would always end at plus nine.

Now in the above example let us do a reverse exercise and *add* one foot to 242 + 09, which will be the first frame you count. The added foot does not end at 243 + 09 as you might have thought but the sixteenth frame is back to 243 + 08.

From this we can establish a formula to apply to our home, or classroom, exercise in taking counts: If you are given the first frame of an optical, *add* the length of

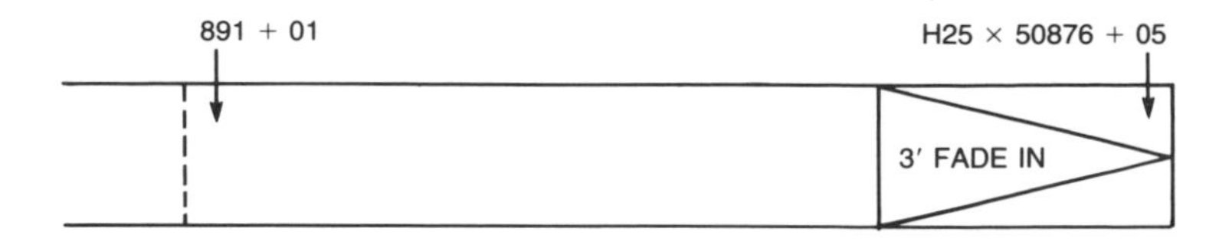

Figure 14–11. Fade in exercise diagram.

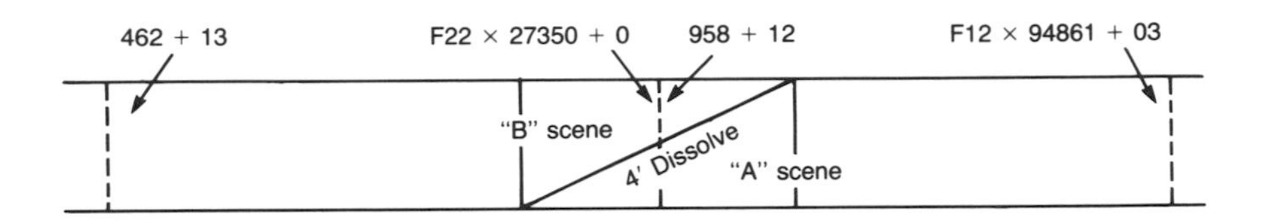

Figure 14–12. Dissolve exercise diagram.

EFFECT NO.	DUPE FROM	START OUT	FULL IN	DUPE TO	EFFECT	LENGTH	REMARKS
203	H25X50875		879 + 04	892	3′	F.I.	CAR to camera A6115 21-1
404	H25X37215	240+09		245	3′	F.O.	KirKwood leaves A4937 48E-4
106A	F12X94860	956+13		962	4′	DISS	Kitchen Scene A9110 14-3
106B	F22X27347		351 + 15	464			Gardener AA1396 15-2
	# 106 AS an INSIDE OPTICAL						
106A	F12X94956	956+13		962	4′	DISS	Kitchen Scene A9110 14-3
106B	F22X27347		351+15	353			GARDENER AA1396 15-2

Figure 14–13. Exercise counts.

that optical *less one frame* for the full in. If you are given the last frame of an optical, *subtract* the length of that optical and *add one frame* for the start out.

Remember that this is only an *exercise* by which to calculate counts without film or synchronizer given certain information. You will *never* have to figure out counts in this manner in any cutting room where you will be able to count the frames on the actual film either by eye or with your sync machine based on the editor's optical marks and your verification of the accurate footage by *blocking in* those marks. However, this exercise does emphasize the correct use of the synchronizer when you are measuring film. After setting your footage counter at 0000, set your framer at 1 so that you will always include the first *frame* as well as the last frame in the measurement of the film.

Also do not confuse this formula with simple arithmetic. Five frames plus 8 frames are still 13 frames, and 2 feet + 09 plus 3 feet + 11 still equals 6 feet + 4, 7 frames minus 3 frames are 4 frames, and 10 feet + 06 minus 4 feet + 15 equals 5 feet + 07.

Now that you understand all this, are you ready for another exercise? Try not to look at Figure 14–13 until you have completed this exercise.

> For the third optical in reel 2, order a three-foot F.I. for a cut in which the first frame is H25X50876 + 05 and the last frame, 891 + 01. Use the customary inside count for this outside optical with one-foot handles.

Refer to Figure 14–11 and enter the counts on your sample count sheet.

Have you completed all the necessary information including filling in the *remarks* column? Now compare your F.I. counts with effect 203 in Figure 14–13. Did you have the correct effect number? Did you incorrectly write *879 + 05* for the full in, or did you remember that when you are adding footage you subtract one frame?

Your final exercise is a little more difficult.

> For the sixth optical in reel 1, order a four-foot dissolve. The first frame of the first cut is F12X94861 + 03 and the last frame is 958 + 12. The first frame of the second cut is F22X27350 + 0 and the last frame is 462 + 13.

The dissolve is diagrammed for you in Figure 14–12. Do these counts *twice,* first as an *outside optical* and then as an *inside* one. Try to do the above exercise before you examine the correct counts in Figure 14–13.

LOPSIDED OPTICAL

By a combination of instinct and perception a good assistant can sometimes detect an error in an optical ordered by the editor. One type of error may occur, for example, after the final cut when it has suddenly been decided instead of a straight cut to dissolve from a night scene in a restaurant to a day scene in an office. The editor hastily draws a diagonal line across the splice for a three-foot dissolve.

As the assistant takes the counts on this optical, she instinctively suspects that the end code number of the restaurant scene, the *A* scene, is dangerously close to the end of the complete take and verifies this in the code book. At the same time the assistant has identified the film and gets the trim roll. She finds the end trim and informs the editor that there are only 12 frames of usable film at the end of the take, not enough film to cover the dissolve. Disregarding any handle at least 12 more frames are needed.

The editor has several alternatives. First the dissolve could begin 12 frames earlier trimming 12 frames off the end of the cut. (If the picture footage is locked in, 12 frames must be added elsewhere.) If, because of dialogue or some other reason, this is impractical, the optical could be reduced from three feet to two feet, or the editor could instruct the assistant to order a *lopsided dissolve,* letting the *A* side fade out before the office scene, the *B* scene, has fully faded in. Depending on the kind of scenes involved, however, this does not always result in an acceptable effect and the final decision may be simply to replace the uncomplicated straight cut.

You have now learned the elementary and most difficult part of opticals. In this chapter you learned the duping process, the identification of the basic effects, and the use of key numbers in the ordering of opticals on count sheets. You are now well prepared to learn more about opticals in the next chapter.

CHAPTER 15

MORE ABOUT OPTICALS

OPTICAL ORDERS

The count sheets are your order for opticals to the optical facility. In addition you should attach a separate order itemizing the number of fades, dissolves, and other effects you are ordering. A copy of this order should be sent to the post-production supervisor. Of course you should keep copies of everything. You also may have to send the optical house three-frame film clips of *every* scene listed on your count sheets so that they can better match the appropriate color. Obviously you retrieve these clips from the trim rolls of the respective scenes just as you did when you had to supply crew members with film clips (see Chapter 11).

Remember that all opticals must use interpositives that are made from the original negatives. Most optical houses insist on making their own interpositives. Since they need the original negatives to accomplish this, send a copy of the count sheets to the negative cutter so that he will know what negatives to send to the optical house. These negatives must be returned to the negative cutter as soon as possible for further use in negative cutting.

If a lab is making the interpositives, you will have to write a lab order. You should write on your order, ''Please make the following paper to paper interpositives'' and list the *dupe from* and *dupe to* key numbers of all the negatives involved. *Paper to paper* written on your order means you are ordering only a particular section of a take. You must also include the slate number, date shot, and negative roll number for each item. Do not neglect to add any necessary instructions such as ''Please send I.P. to ____________ Optical Company.'' When the lab makes the I.P.s, the negative cutter receives a copy of the lab order instead of the count sheets and sends the appropriate negatives to the lab. The lab sends the I.P.s to the optical house and returns the negative to the negative cutter.

REGISTRATION INTERPOSITIVES

Registration I.P.s are specially processed by step printing, or printing frame by frame, for maximum stability and to obtain optimum quality in particular effects. Whenever an optical house has work with various mattes, process plates, titles, and other similar effects, registrations must be ordered instead of ordinary I.P.s. You may not have to make specific orders for registrations if your optical house, like some in the Hollywood area, automatically register *all* I.P.s they make regardless of the effect.

GENERATIONS

Beginning with the original negative as a *first generation,* an internegative derived from that original negative would be termed a *second generation.* A new internegative obtained from second generation material would then be *third generation,* as in the main title situation that is cited later in this chapter.

MISCELLANEOUS EFFECTS

Now that you understand how to do counts for fades and dissolves you should have no problem with such opticals as wipes and flips that may only vary in length or require some additional information. Wipes and flips have the same start out and full in counts as do dissolves.

In addition to these basic opticals, there are a number of other effects that, like straight dupes, do not fall into the same category but are still included in your counts. In some cases as in a freeze frame, move-in, or blowup the start out and full in columns on your count sheets may be used advantageously provided you clearly indicate what you want. Sometimes one of these effects may begin or end with a fade, dissolve, or other optical. Figure 15–1 illustrates descriptions and counts of some of the most commonly used *miscellaneous* effects.

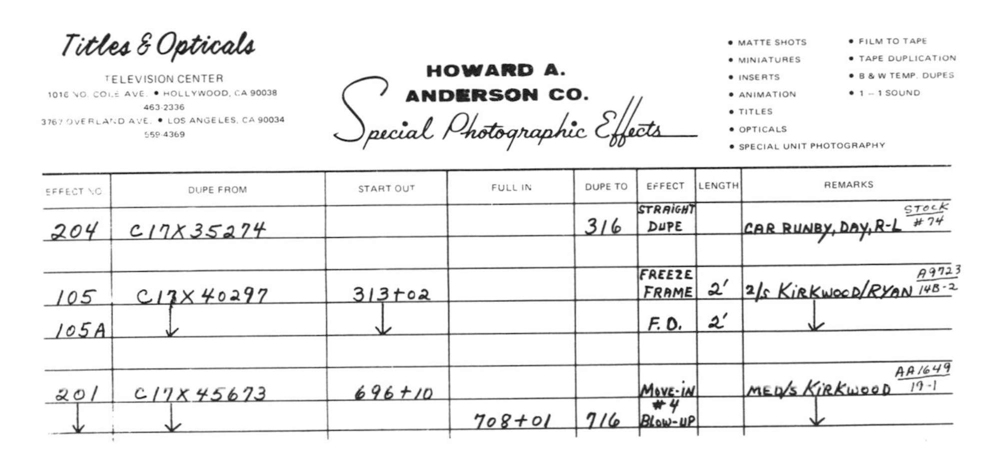
Titles & Opticals
TELEVISION CENTER
1016 NO. COLE AVE. • HOLLYWOOD, CA 90038
463-2336
3767 OVERLAND AVE. • LOS ANGELES, CA 90034
559-4369

HOWARD A. ANDERSON CO.
Special Photographic Effects

• MATTE SHOTS
• MINIATURES
• INSERTS
• ANIMATION
• TITLES
• OPTICALS
• SPECIAL UNIT PHOTOGRAPHY
• FILM TO TAPE
• TAPE DUPLICATION
• B & W TEMP. DUPES
• 1 – 1 SOUND

EFFECT NO.	DUPE FROM	START OUT	FULL IN	DUPE TO	EFFECT	LENGTH	REMARKS
204	C17X35274			316	STRAIGHT DUPE		CAR RUNBY, DAY, R-L STOCK #74
105	C17X40297	313+02			FREEZE FRAME	2′	2/S KIRKWOOD/RYAN A9723 14B-2
105A	↓	↓			F.O.	2′	↓
201	C17X45673	696+10			MOVE-IN #4		MED/S KIRKWOOD AA1649 19-1
↓	↓		708+01	716	BLOW-UP		↓

Figure 15–1. Counts for miscellaneous effects.

Freeze Frame

These are also called *hold frames* or *stop frames.* To create one the editor selects a particular frame to repeat for any desired length. This optical is used for dramatic or comedic effect and is used so often at the fade out of an act in episodic television that it was one of the four major *missing banners* you gave your editor (see Chapter 6). Effect 105 in Figure 15–1 orders a freeze frame using the *start out* column to indicate the exact frame to begin the freeze. Note that the entire cut is being ordered; you should not order inside opticals for freeze frames.

Flopover

A *flopover,* or *flop,* not to be confused with the flip that was described in Chapter 14, changes the cell side of the film to the viewed emulsion side so that the emulsion side becomes the new cell side. It is usually used to correct misdirection. If a performer enters screen left, a flop will result in the performer entering screen right. The film should be carefully examined for any possible incongruity. For example if there is a sign in the background, the writing would be reversed. Unless you are able to *blowup* and remove the sign, this flopover would be unworkable.

Your editor can flop a cut in the work print, but this is only temporary. On the screen it will appear out of focus. The original negative cannot be flopped; it must be done optically and ordered on your count sheets.

Tail-to-Head Reversal

In this effect the tail, or end, of the cut becomes the head and the head, the tail. Thus a car coming toward camera and stopping can become a car starting and backing away from camera. Other elements in the scene should also be carefully examined for anything that might become discrepancies.

Years ago such an effect was once ordered on a scene of a train coming toward camera so that it would appear to be backing up. No one noticed until after the picture was released into theaters that the locomotive smoke was going down instead of up as a result of a tail-to-head reversal.

Skip Frame

A *skip frame,* or *skip print,* is created by the elimination of frames within a cut to speed up the action. To save the expense of an optical while he is experimenting, your editor may ask you to omit every other frame of the work print so that he can view the result. As you eliminate each frame temporarily attach the retained frames to each other with small pieces of tape on the cell side. When the cut is complete, use as large a piece of tape as possible for the emulsion side; you may have to do it in two or three sections depending upon the length. Then remove the small pieces of tape on the cell side and repeat the taping.

As you remove each frame to be skipped, carefully identify it in case you have to replace it before a reprint of the scene can be ordered. Identify each frame by inscribing a key number reference with your white grease pencil. For example, if the first good frame is at H35X15904 + 13, you would write *904 + 14* on the first frame you take out, *905 + 0* on the second skip frame, and so on. A skip frame segment of film can be devised in various ways such as skipping every third frame.

Speeding up the action would not have to be done in editing if the scene had been shot *undercranked,* which is decreasing the normal camera speed of 24 frames per second. When you are ordering your optical for a skip frame, be careful to identify the correct frame in the handle to begin skipping so that you will have the correct good frames to cut in. Remember that handles are for possible use should the editor decide to adjust the length of the optical.

Double Printing

Double printing, or *multiple printing,* is created by slowing down the action such as when a scene is shot *overcranked,* increasing the normal camera speed. You cannot experiment with this in the editing room as you did with skip printing. It must be ordered as an optical with the instruction to print *every frame twice, every other frame twice,* and so forth.

Blowup

Your optical service will provide your editor with a *blowup chart,* having nearly a dozen fields, or sizes, of blowups from which he can make a selection by number. Number 1 on the chart represents normal aperture. A number 2 or 3 field, which is a slight blowup, could eliminate a disturbing microphone hanging down into a scene.

The editor can thus create a big head closeup from a medium closeup by ordering a 6 or 7 blowup. The larger the blowup, however, the more danger there is of the effect being grainy. To avoid this, you might order the blowup "as large as possible *without grain,*" leaving the specific chart number to the expert judgment of the optical technician.

Move-ins and Repositions

A *move-in* is self descriptive. It is optically moving in on a person or object. A continuous move-in ending in a dissolve to another moving shot can sometimes be ordered to improve a scene transition. As with a blowup a move-in can at some point become grainy. This should be avoided.

A *reposition* is created by optically adjusting the original angle of a shot. Depending on the material, there is a limit to how much of an adjustment can be made.

Move-ins and repositions, like blowups, are ordered for various reasons such as to create an angle that was not originally shot, to improve on an angle that was shot, to eliminate something or someone in a scene, or to heighten dramatic effect.

You must order these effects as *outside* opticals, indicating the exact frames at which you wish the effect to begin and end. Blowups are often used in conjunction with repositions or move-ins (effect 201 in Figure 15–1) and sometimes all three are used in one effect. For example, a move-in on an individual in a scene with other people can be repositioned so that the object will be center screen by the time the numbered blowup requested has been reached.

Superimposures and Montage

A *superimposure,* or *super,* is the overlapping of two or more scenes which, unlike those in a dissolve, are simultaneously of constant relative strength. Uncomplicated superimposures (such as no more than two scenes on screen concurrently even though there may be a series of such scenes) can be ordered on the count sheets if you provide all the necessary information described below. Any sequence of superimposures is usually referred to as a *montage* when it includes many cuts that are constantly changing, dissolving, and conveying a transition in time and events in a relative short screening time. One or more supers will usually be found in most montages.

Multiple supers overlapping one another, some with fades, dissolves, or other effects, are difficult to outline clearly on a count sheet. It is much simpler to present the counts to the optical house in the form of either film or *templates,* the use of leader. This eliminates a lot of paper work and more importantly decreases the chances of misinterpretation. Instead of the counts being submitted to the optical people on count sheets, you will be giving them a replica of the film involved on which all the necessary information will be inscribed.

If you are going to send them B&W reversal film you must first send the original color prints of all the cuts involved to a reversal house. Some of the prints will have to be removed from the cut reels. Always splice in temporary leaders, in sync of course, and mark on the leaders what you have done with the color. Do not do this unless you receive your editor's approval. When you receive the reversals, replace the color prints. The B&W reversals will contain all the code and key numbers as they were on the color prints.

As an example, imagine that you are ordering a montage consisting of five different scenes on four rolls, each on a core. Place a short piece of leader on each roll, identifying them as supers *A, B, C,* and *D* (see Figure 15–2). Super *A* is the scene in the picture preceding the montage and continuing into it. Super *B* has two scenes; the first scene fades in supering over *A* then dissolves to a second scene as super *C* fades in. As super *D* fades in, there are four scenes superimposed at the same time before *A* fades out.

As each super appears, you must establish a sync with a key number at the zero frame of another super. Thus in Figure 15–2 using only the last three digits of a key number, *B* super syncs to *A*'s key number, 296 + 0. When *B* dissolves to another scene, a new sync has to be designated with *A,* 327 + 0. *C*'s sync is with *B*'s key number, 716 + 0, as is *D*'s at 731 + 0.

There is no particular rule as to which super you use as a sync. In Figure 15–2 super *B* has no choice; it must use *A* for sync because that is for a while the only other super. However, super *C* could have still used *A*'s key number instead of super *B*'s, and super *D* could have synced to *C* instead of *B*. Thus there are instances when you have choices you can make in the matter of sync.

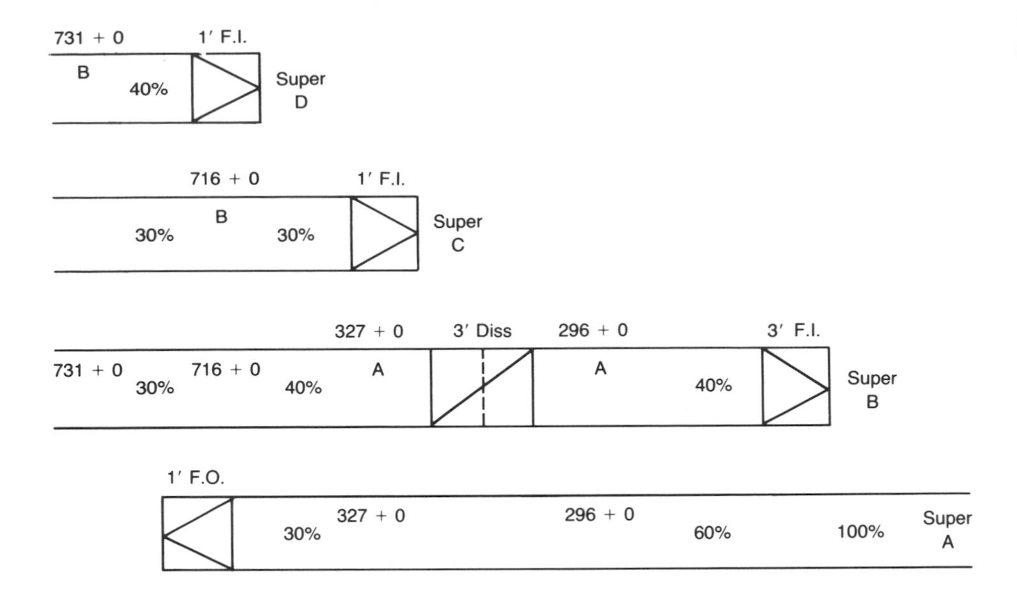

Figure 15–2. Superimposures.

You must make certain that the fades, dissolves, and other effects within the montage are closed in to accurate footages and clearly identified as to the optical effects desired.

Your editor must specify *image strength* percentages of each super each time one begins or ends. Of course, they must add up to no more or less than 100 percent at all times. Super *A,* since it precedes the montage and is on screen alone, begins at 100 percent. As soon as super *B* begins to appear *A* must share the 100 percent. Whenever there is more than one image on the screen, the *full* image strength must be shared. A single scene at full strength usually begins and ends a montage but a single scene may be on screen anytime *within* a montage, depending on how it has been edited.

How does the editor arrive at these percentages? First he decides which super has the most importance to the story as compared to the corresponding super(s) and assigns the higher percentages accordingly. Then the editor tries to envision the effect using any past experience with superimposures and his expertise and creative judgment. Since superimposures and montages must be ordered as a continuous *outside* optical, the completed effect will be sent to you in *one* section and unfortunately, usually because of the percentages, may not fulfill your editor's expectations. The entire montage may have to be redone several times before everyone is satisfied with it.

In some situations you may be permitted to send the marked color prints to the optical department and will not have to order reversals. If for some reason neither color prints or reversals can be sent, you will have to use *templates.* This is simply using *leader* instead of film. Everything pertinent on the original prints must be duplicated exactly: the exact length of each cut, the identification of each cut with the full key number marked at the correct frame at least once either at the beginning of the cut or at the full in of an optical, and the inscription of all the other information described above.

Finally whether you send the optical facility film or a template, make certain you have all the information duplicated on the film you have retained or in some other manner so that you can refer to it should any questions arise.

SCREEN CREDITS

There does not appear to be any standard procedure as to whose credits appear in the main title, opening credits, or end credits of episodic series (refer to Chapter 13). Certainly there seems to be a good deal of variation. The decision as to who gets what where is the prerogative of the production company in collaboration with the network. The decision, however, must take into consideration any contractual agreements and the length requirements as set forth in the network format (see Figure 13–3).

Generally you can assume that the series title and the stars and regular, or *running,* performers will be credited in the main title along with the writers or creators of the series. The executive producer credit may also be in the main title, but it is often superimposed over the fade out of the picture.

You might expect the opening credits to begin with the episode title if it is credited at all. Most series do not list this *working* title. The opening credits will usually include the guest performers, writer(s), and director of the episode.

All other credits including those of director of photography (DP), art director, and editor are usually in the end credits with other crew and production credits. Unfortunately, I am sorry to say, assistant editors are rarely ever given screen credit on episodic series.

Main Title

The counts for a main title may have to be ordered with film or template as with a montage if it is involved or complicated. When it is finalized, it is in the form of a single internegative that may never be cut because this main title must be used for every episode of your series. Therefore a *master* I.P. must be ordered from the interneg to protect it just as you preserve original stock negative. Each time a main title is needed for an episode, a *third generation* internegative is ordered from the master I.P. and a dupe print from the interneg. Of course, you replace the main title leader in your work print with this dupe print, and the interneg is used by the negative cutter to cut into the negative reel.

Later in the season should there be any change in the main title such as in the personnel credited, requiring a new title card, a new internegative is made and from that a new master interpositive.

Character Deletion Report

As soon as is possible your editor should give you permission to send a character deletion report on the episode to the responsible party, usually the producer or the legal department of the company. The report is required when all scenes with a particular actor or actress have been deleted from the final picture. This usually occurs when footage must be lost and the deleted footage is considered less germane to telling the story. On the report list your production, production number, the performer's name, and the name of the character he is portraying so the person's screen credits will be deleted from the title/credits sheets. This will eliminate the unnecessary payment of residuals to that performer.

Title/credits sheets for opening credits and end credits must be approved by the producer and copies sent to your editor and the optical house or title department before work on these credits can begin. The sooner you send in your deletions report, the sooner the lists can be sent out and work can proceed. Usually there are no deletions and the assistant must so advise someone by phone. Often an assistant forgets to make the call for a day or two creating unnecessary delay. It might be a good policy to always send a report, either listing the deletions or marking the report as having *none.* If you do not receive credit sheets in reasonable time, you should remind the proper individual.

Opening Credits and End Credits

Opening credits cannot be timed out on the picture nor counts taken until your editor receives a copy of the approved credits that are listed by *card* numbers. Credits and titles are usually printed on *cards* that are individually photographed and superimposed over a selected background. Thus, the numbered cards are used on the work print and count sheets as identifications of the particular credits.

Usually opening credits are superimposed over the action at or near the opening of the episode. The credits sheet lists the order of the credits. If it is received in time, the editor lines or marks the film where he intends to place the credits before the final cut so the producer will be informed. Once he has been advised of any network or performer contract requirements regarding screen credits, the editor determines how much footage to assign to each card so that it might be easily read, depending on the number of names on each card, whether the cards should fade in and out or pop in and out, the amount of footage separation between the cards, and where to position the titles within the picture frame.

While your editor has considerable latitude with the opening credits, the end credits are limited to the footage dictated by the network. In your sample network format (see Figure 13–3) 40 feet + 08 is required. The 5 feet + 08 *logo* that follows the end credits and ends the film may come to you already attached to the end of the end credits. It is a filmic symbol of your series' producing company and, like the main title and bumpers, is repeated in each episode.

It is the assistant's responsibility to check credits at each stage for any errors, the most common being the misspelling of names. Other errors might be the improper inclusion or deletion of a credit, and if character names are listed with the performers, whether they have been properly assigned. Since it all begins with the credit sheets, both you and your editor should carefully check them and immediately report any error. Check the corrected sheets just as carefully.

Even though the title/credits sheets are correct errors can still be made by those who are responsible for the cards. Therefore when you receive the filmed credits, before you cut them into the picture both you and your editor should check them for any further mistakes particularly misspellings.

COLOR REVERSAL INTERNEGATIVE—(CRI)

This was an editing room term for what most film labs used to call a *color reversal intermediate.* The past tense is used because since late in 1987 CRIs can no longer be ordered, at least not in Hollywood. By mutual agreement Eastman ceased to provide the necessary stock and Consolidated Film Industries (CFI), which had the required machinery for processing, suspended that particular operation. Why then should it be discussed in this text? There are two reasons: (1) because it was an important phase in past editing and you should be knowledgeable about it; (2) because somewhere outside of Hollywood there may still be some available CRI stock and a lab that can process it.

The color reversal internegative replaced the interpositive and internegative in the duping process (see Figure 15–3). It eliminated one step in the process. Therefore it took less time to process. Standard time for an I.P., interneg, and dupe print is about five days. The CRI and dupe print took only about three days. It had some disadvantages, however.

The color quality was generally inferior to that of an internegative. In addition the length of fades, dissolves, and similar opticals were erratic when CRIs were used so that adjustments had to be made in the optical processing to approximate the desired length. CRIs were never used for complicated opticals or titles. And since the CRI was cut into the negative reel by the negative cutter, should the optical have to be redone for any reason or should the same material be used again, the original negative had to be used to make another CRI. This was a disadvantage because the less the original negative is handled the safer it remains.

Considering these disadvantages why did any company ever consider ordering CRIs? In addition to saving time CRIs saved money. Before an assistant ordered any opticals or other dupe material he first determined whether his company wanted him to order I.P.s or CRIs. CRIs were never used on major films, theatrical or long-form

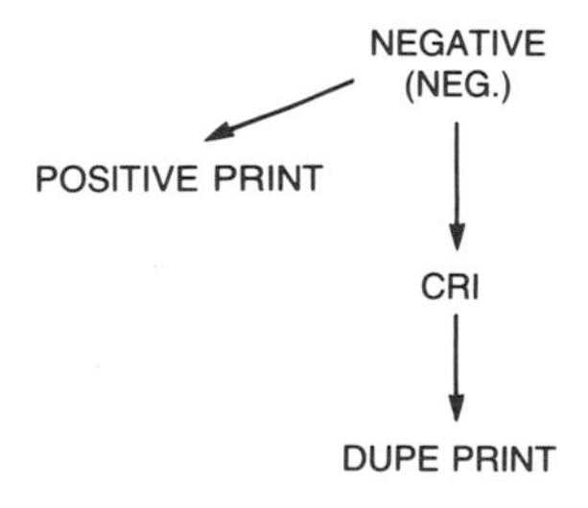

Figure 15–3. Color reversal internegative (CRI) chart.

television. They were ordered on some low budget films and on many TV series. When the assistant received the dupe prints from the CRIs, he would cut them into the work print thereby replacing the prints from the original negatives. The CRIs would go to the negative cutter.

Whenever you, the assistant, have the slightest doubt about how to write up counts on a particular optical or what should be ordered for that optical, *always ask* your editor or a knowledgeable fellow assistant. Actually, the best person to ask is your contact person at the optical house. He is usually most helpful and will appreciate your concern. In fact, should your editor be in doubt about the kind of optical that will achieve a desired effect, an optical person will often come to your cutting room to advise him. At the same time you can ask this person how to order the optical. Always remember that an error in the counts that results in having to remake an optical can cost your company a lot of money.

EXCEPTIONS

Screen Credits

On features and most longer TV projects main titles will generally include those individuals listed in opening credits on episodic series; the chief producing credits; and such technical credits as director of photography, production designer, composer, and editor. These are sometimes referred to as *up front* credits. Some of these technical credits, including the editor, may receive credit on feature advertisements and billboards. Assistant and apprentice editors do appear in the end credits, particularly in features.

A&B Printing

When A&B rolls are used for negative cutting (see Chapter 16) and subsequent printing, fades and dissolves are automatically done without any need of count sheets. The work print is sent to the negative cutter who uses the editor's optical markings to prepare the negative so that these opticals are made simultaneously with the processing of the composite print. All other opticals must be ordered from an optical company.

The negative cutter uses key numbers to match precisely and to cut the negative exactly as the work print has been edited. Standard negative cutting is building *one* negative reel or roll to correspond to each work print reel. A&B cutting is using *two* rolls and on 16mm cutting in the appropriate negative alternately on one roll and then the other *after each cut.* This is called *checkerboard* cutting. On 35mm film the change from one roll to the other is only done at each fade in, fade out, or dissolve.

Each lab has maximum lengths for those opticals that can be effectively processed in this manner. Each lab also has a minimum length that must exist between those opticals that closely follow one another. Therefore you and your editor should be advised of your lab's requirements.

A&B printing is only occasionally applied to 35mm film usually when a company wants to save some optical expense. It is more generally used on 16mm products. Opticals executed on A&B rolls are of much lower quality than those done by an optical printer, and the obvious disadvantage is that the completed effects are not seen until they are viewed in the composite print when it may be too late to make any corrections.

The Hybrid System

On a show that is shot on film and edited completely in post-production on film, negative is cut to match the cut picture (see Chapter 16). After the composite print has been approved, a film-to-tape transfer to a one-inch videotape *master* is made at a video facility in a *telecine* session in which a camera-projector combination electronically photographs the negative image. Videotape copies are then made from this master for television viewing.

As indicated in Chapter 1 most television shows that are edited *on tape* are still shot on film. In this situation the negative is *never printed* but instead is immediately transferred to tape. After the picture has been videotape edited, it goes to a video facility for an *on-line* session. During on-line sessions the titles and required optical effects are created and a one-inch videotape master is produced from which three ¾-inch cassettes and three 2-inch 24-track masters containing the *A* (production) track are made in a transfer session called a *lay down.* These are given to the sound effects, dialogue, and music editors. When their work is done, it is *mixed* or *dubbed* (see Chapter 16) onto one of the 24-track masters and, in a transfer session called a *lay back,* married to the one-inch master in sync with the video. Meanwhile the original film negative is *never cut.*

There is also that which I call a *hybrid system* using both film and videotape techniques to complete a project. During the past couple of years a few major television companies have begun creating opticals and titles electronically for some of their shows that have otherwise been completely edited on film. Several different videotape facilities have developed this procedure. Joseph Dervin, Jr., Vice President of Post-Production at Aaron Spelling Productions, explained the reasons for using this hybrid system on their *Dynasty* series, which is shot and edited on film:

> It is a process that marries some of the best currently available video processes with some of the film techniques that we've been using for years and gives us a better vehicle for delivering shows produced specifically for television.

Dervin then explained why his company had not gone completely over to videotape:

> . . . unfortunately, by making the decision to complete a show using electronic post production, you are buying yourself some negative aspects. The first of those is the master that you're going to keep on the finished show, the only element that you're going to have for the future to cover all the current and foreseeable world-wide markets for that piece of entertainment is a one-inch videotape . . . in the current 525-line NTSC (National Television Standard Committee) format which is our standard in the United States. Nobody knows how long that piece of videotape will last. Nobody will at this point warrantee that in fact in ten years you can pull that one-inch videotape out of the vault and successfully make a good copy.

> Secondly, because the master is in the NTSC 525-line format, the only currently available method of delivering to foreign companies which use a different format, either PAL or SECAM both of which is 625 lines, is by going to a process called Standards Conversion which is basically a black box that converts 525 lines, 60-cycle (hertz) information to 625 lines, 50-cycle (hertz) information in either the PAL or SECAM format. It's a process that delivers to the foreign customer an element that is not as high in quality as a PAL or SECAM master made directly from a print or cut negative.

It should be noted here that most foreign countries use the PAL format while France, Russia, and more than twenty other countries edit in PAL and transmit in SECAM. PAL means *phase alternate lines.* A loose Hollywood rendition is *peace at last.* In Brazil the format is labeled PAL–M; the Hollywood translation is *peace at last–maybe* or *pay a little more.* SECAM is *sequential couleur á memoire* (English translation: *Sequential color with memory.* Hollywood translation: *System essentially contrary to American methods*).

Dervin's strong concern regarding foreign distribution is better understood when we learn that *Dynasty* is sent to 95 countries. He is trying to deliver the best possible quality product and at the same time is trying to protect his future ability to service the foreign market even if the format transfer requirements change in the future. Dervin believes that by using the hybrid system he is achieving the best of both film and electronics because, (1) it provides him with a cut negative that will be his archival master in the vaults forever, and (2) not only can he transfer to NTSC videotape but he can also transfer directly from his *film master* to PAL videotape.

This procedure retains all other aspects of film editing and post-production, including the cutting of the negative to match the picture. The only missing elements in the cut negative are the completed opticals, titles, and production stock footage. Thus the company continues to use the same film equipment, personnel, and procedures—the exception being the manner of ordering opticals.

Sharon Bernard, one of the three assistant editors on *Dynasty,* explained how opticals and titles are ordered. Instead of count sheets, a similar form called a *log sheet* is used. *Everything* except cuts is individually logged. This includes opticals, titles, credits, stock, and all the format material. Even the *black leader* where it is specified in the format is logged. Each of these items is called an *event.*

When she begins to log each reel, Bernard sets the footage counter on her synchronizer at 000 + 0 on the first frame of picture. Each reel is logged on separate sheets identified with the reel number. She indicates each event in a reel by *footage plus frames.* Key numbers are not used nor are I.P.s or internegs ordered. The only exceptions are when outside stock is used or a particular optical is required that is more effectively created on film than on videotape. An example of the latter is a shimmer dissolve flashback to a vignette commonly used on *Dynasty.*

Production stock is mounted on reels and used as a *library* reference. *Dynasty* has nine reels of stock, each coded from the first frame of picture as the zero frame. Thus, reel 1 begins with 1000 + 0; reel 2, with 2000 + 0; and so forth. The corresponding negative rolls are filed at the videotape facility. When her editor selects a piece of stock from a library reel, Bernard orders a B&W reversal to be cut into the picture. The reversal will have the footage code as a reference for the videotape personnel.

When editing on a show is completed, the negative cutter matches negative to the work print using *A* and *B* rolls as previously described. Most of the picture is on the *A* roll, including the *A* side of opticals, while the *B* roll is used primarily for the *B* side of opticals. At the first cut following the optical the negative cutter immediately switches back to the *A* roll.

The A&B negative rolls are sent to the videotape company where they are transferred to two one-inch videotapes and at the same time color corrected in a telecine session. "When you transfer to videotape from a timed, color-corrected, low-contrast print," says Dervin, "you still lose some details—some visual information. . . . By going directly from the *negative* to videotape, it's no longer lost. . . . This process allows us to deliver a better looking element for broadcast in the United States by the network."

The dubbed sound is transferred from a dubbed master and the one-inch videotapes of the A&B negative rolls are put through an on-line session during which all opticals, titles, and production footage stock are added. The result is a single, one-inch videotape master.*

More Optical Exceptions

Since an episode of a series is in the cutting room from first to final cut for such a short time, normally about two weeks, it is logical to order the majority of the opticals at one time when the picture is locked in and when there is slight possibility of any editing changes that might affect the optical counts. Any later alterations in opticals could possibly endanger the air date.

On longer projects, TV movies and features that might be in editing for from four months to a year, the assistant often has to order opticals periodically. Some of them may have to be reordered due to editing changes, but by the time the picture is turned over most of the opticals will have already been cut in. These opticals are ordered in such a disorganized fashion that the usual numbering system on the count sheets might not be effective. Instead of identifying the optical by reel number and placement in the reel, you might consider using the general scene number.

If you have a sufficient number of such opticals, you might consider coding the dupe prints in sync with the tracks. It will help your editor when he is re-editing. Devise some system with the letters of the code numbers or, if you are coding on an Acmade, with color to specifically identify them as opticals.

Recently in a few companies, evidently with the approval of their optical facilities, some assistants have not been filling in the *start printing,* or *dupe from,* columns for fade ins and for the *B* sides of dissolves, wipes, flips, and so forth; nor do they complete *stop printing,* or *dupe to,* columns for fade outs and the *A* sides of those opticals. Having the full in and start out counts and the lengths of the opticals is sufficient

*For complete information on all aspects of videotape and electronic post-production I recommend *Electronic Post-Production and Videotape Editing* by Arthur Schneider, A.C.E.

for the technicians operating the optical printers. When the optical house is using your count sheets to make the interpositives, however, which requires all the beginning and end key numbers, these assistants are relying on the lineup person to provide the missing counts. And when the assistants have to order the interpositives separately from the lab, they have to include all the key numbers. So why not include them on the count sheets, serving as an excellent double check, for themselves, the optical department, and the lab? Should an assistant err on an I.P. he has ordered, such a double check could prevent a possibly fateful delay. Leaving those columns blank is not that much of a time saver, and how can one take pride in one's work if that work is incomplete?

Would you like to order your opticals without taking *any* counts? Some independent productions, usually nonunion or student, make arrangements that, if they send the color work print to the optical house, the line-up person will log the optical counts from the film thus eliminating the need for any count sheets from the cutting room. It also places the full responsibility for the counts on the optical company. Before sending reels, instead of counts, the assistant must still go through the picture, confirm the length of the effects, and close them in, clearly identifying each effect on the film and numbering it.

No count sheets sounds wonderful, doesn't it? What then are the disadvantages? First neither an editor nor a company likes to release the work picture to anyone unnecessarily because the picture might be needed for an unexpected editing change or screening, or the negative cutter may need it. Also if there is no double check, the line-up person, being unfamiliar with the picture and your cutting room procedures, may misinterpret some opticals and cause unfortunate delays. In addition the optical company could realistically add expenses for undertaking the assistant's responsibility. Finally the inference made, when this service is requested, is that the request is made because "evidently no one in the cutting room knows how to write up count sheets." Do you want that kind of reputation in the industry?

CHAPTER 16

FINALIZING THE FILM

Once she turns over the picture to M&E and orders all opticals, title, and credits, the assistant may have a couple days before she receives the optical dupe prints and begins dailies on the next episode. There will be time for miscellaneous chores some of which are described below.

STOCK REPORTS

Two different reports relating to stock footage may be required. The first is for all *production material* you red lined on this episode that should be saved for stock for your series including any additional scenes your editor may have selected. You should list the series name, episode production number, slate, beginning and end keys, and a brief description and whether you ordered an interpositive for use for future dupes. The report goes to the film librarian who advises the negative cutter to place the negatives under the series special stock number.

A second stock report is required for *all outside film library stock* actually used in your episode. This report will note the series name, episode production number, name of film library, description of stock used, can and roll number, and beginning and end key numbers to indicate total footage of the film used in the picture. The latter figures are extremely important because they determine the cost to your company for use of that stock (currently for color: $30.–$35. a foot with a 10-foot minimum). This report is sent to your film library for proper disposition.

PREPARING FOR THE NEXT EPISODE

You now have time to prepare the daily leaders for the next episode. Be sure to save the leaders from the first episode. Now you need only remove the tape identifying that episode and replace it with new tape and the new production number. Since the first episode used *A* code numbers, you might begin these new dailies with *B*-1000.

Check your supplies and those of your editor and replenish anything that is depleted. It is still early in the series and if you properly supplied your room on that first day, you should not need anything yet. But some items diminish faster than expected. Do you have enough empty reels and trim boxes? Is all your equipment in good working condition—such as the butt splicers, the moviola, rewinds, or lightboard? Is your editor still supplied with good grease pencils, note pads, and paper clips? Should anything require servicing this is a good time to have it done.

SOUND EFFECTS

Sound effects editing involves using additional production tracks and substituting other production tracks with sounds from a sound effects library or with ADR and foley—all for the purpose of implementing and refining that which was originally recorded. *Sound editors* accomplish this in cutting rooms usually with take-up moviolas (see Figure 11–5) and sometimes using headphones. If they are given assignments other than sound effects, then they may also be called ADR, foley, or dialogue editors.

ADR (*Automatic* or *Automated Dialogue Rerecording* or *Replacement*) is using an automatic electronic system to replace certain dialogue in the picture that has to be redone by the performers usually because of disruptive background noises, erroneous or unclear dialogue, or poor delivery performance. Before this was done electronically it was accomplished by joining the beginning and end of a section of dialogue track to be rerecorded, forming a *loop,* and therefore the procedure was, and sometimes still is, called *looping.*

The ADR stage has a glass-enclosed booth with an electronic console for the sound technician, and at rear center of the stage and high above it is the projection booth. On the stage itself there is a speaker's stand to hold the script lines to be rerecorded and a set of headphones for the actor or actress doing the ADR. The ADR editor sits nearby at a panel with an intercom to the technician and projectionist.

The picture and track of the segment in question must be repeated a number of times so the performer can review the material and redo the dialogue to match the lip movements of the picture image. The ADR editor has provided a *cue sheet* listing footages in the reels that locate each piece of dialogue to be rerecorded. The scene is projected on a large theater-scale screen, and the performer tries to lip sync the new reading. Three short *beeps* before the first word of each short piece of dialogue are electronically injected to cue the actor when to begin. Some actors are excellent at lip-syncing and few takes are necessary. Others have a difficult time and have to do several takes before the ADR editor is satisfied that the best possible reading with the closest sync has been recorded.

Afterward in her cutting room the ADR editor rechecks the selected takes to the picture and makes any necessary adjustments for the best sync. In some cases it may mean a great deal of meticulous work eliminating or adding a sprocket or two throughout a reading, or the ADR editor may have to use a word from an *outtake* to satisfactorily complete a line.

Foley is the process of creating sound effects that did not record satisfactorily during production and are not available in the sound effects library. They are recreated in sync with the film on a sound stage. A major function is the recording of footsteps and the *foley stage* contains every type of surface imaginable for this purpose such as concrete, sand, cobblestone, and wood. Footwear duplicating or similar to that worn by the performers in the film should be used. In some companies the foley editor or other sound effects personnel try to reproduce the footsteps, but I believe best results are gotten when *dancers* are hired. They can more quickly and more precisely imitate the rhythm and timing of different manners of walking.

Foley is not limited to footsteps; a wide variety of other effects can be created on the foley stage. For example, I once observed a foley session when a stunt man jumped from a 20-foot ladder onto a mockup of a car so that the sound of his hitting the car top could be used in a police show.

The foley editor must prepare for each session. She must make up cue sheets indicating where the foleying is required and must see to it that any people or items needed are on the stage.

The *dialogue editor* cleans out usable production dialogue by removing it from the work track to a dialogue 1 reel or unit. The dialogue editor also eliminates any disturbing sounds between words and tries to find a consistent ambience, or background sound, for all the angles used in a particular scene. Sometimes depending on the amount of work involved and the time schedule, dialogue and ADR will be done by one sound editor.

Besides the above-mentioned editors, the sound effects department assigns a different sound editor to each of the five reels. This is necessary because of the limited time they have, only five working days, before *dubbing*. Dubbing refers here to the final editing phase in which all sound effects including ADR, foley, dialogue, music, and retained production tracks are combined to form one composite track. It is also referred to as a *mix*.

Units or reels of track must be built by each sound editor for each of the reels. It is not unusual to have more than 10 units of effects for a single reel. The more effects, foley, and ADR needed in a particular reel, the more units it may require and the more difficult it will be to make the dubbing schedule.

Some sound effects personnel may come to your cutting room with a question regarding the code sheets, or they may request the trims of certain work tracks so they may extend the production effects or be able to make a loop. Try to help them as much as possible. It will be beneficial to your project.

MUSIC

Only one music editor is assigned to your series. The music editor provides the composer with precise timings of the music cues throughout the picture, along with action and dialogue descriptions. When the music is composed, there is a scoring, or recording, session. Then the music editor will build music units for the dubbing session.

The music editor may ask your editor to retime the cuts on a music number. This is sometimes required when the music by which your editor originally cut the number has been replaced with new music. The retiming adjusts the slight change in the music beat, shifting a frame or more from one side of a cut to another. You can help your editor by finding the frames needed in the trims.

EDITING CHANGES IN DUPES

Post-production supervisors try to discourage producers from making editing changes after the picture has been turned over. Such changes can be particularly devastating on a series where there is normally tremendous pressure to make dubbing and air dates. Sometimes changes are unavoidable when, after a late network screening, a standards-and-practices change is required.

Bear in mind that the total action footage has to be maintained. Removing or adding something might appear quite simple but it could conceivably compel the editor to make up that footage in two or three other areas, possibly in different reels, and that might multiply with additional changes.

Sound effects and music reels must also reflect exact changes. How should you correlate the changes your editor has made in the work print with their reels? If an addition of a cut or extension of a cut is only a couple of feet or a few frames, you can duplicate the added footage in the dupes by using leader, inscribing the code number at the proper sync and a description of the action. If more footage is involved, order B&W reversals of the addition as well as track transfers for your own and music's dupe tracks. The original track of the added footage goes into the work track that sound effects has.

Merely adding or deleting any footage is not sufficient. You must keep a record of the amount of footage involved, the reel and footage in that reel at which the change was made, and the date of the change. This is a continuous record with copies, including any later changes going to the sound and music editors.

MEASURING FOOTAGES

Measuring footages for music and effects is different than measuring action footage. The latter is measured from the first frame of picture beginning with 000 + 01. Measured footages for music and effects must begin at the 12-foot academy start. This means measuring *everything*—action and format material.

There is a curious disagreement between different M&E departments and also between the editors within those departments about whether to set the start mark on the synchronizer counter at 000 + 0 or + 01. In my opinion it should be set at + 01. But try to learn which is the method preferred by your department. If you are unable to do so, the choice is yours. In either case, indicate in writing to M&E how you are starting to avoid any confusion or error.

Hypothesize that your editor has added a 5 foot + 03 cut at 225 feet + 07 in reel 2 and has deleted equal footage at 317 feet + 10 in the same reel. When you make the changes in a dupe and since you will be taking footages at the same time, splice in the 5 feet + 03 additional cut on the *left* side of the synchronizer. The footage counter will register the added footage, and you will have a true footage down to your next change—the deletion at 317 feet + 10. Deletions should be made in the same manner on the left of the synchronizer.

At the head of an added cut or at a splice place a small piece of tape on the left side of the film and inscribe on it the counter footage and amount of footage added or deleted. With this and the record you give them the music and sound editors should have no doubts regarding the changes.

If these editing changes are made after M&E units have been built, the corresponding changes must be made in *every* unit. The task of making all the reels conform becomes more arduous and the risk of error increases. A footage error in any reel could result in a costly delay on the dubbing stage. With so many reels involved the assistants in M&E will most likely be assigned the work. Your record should always note whether you or the other editors actually cut in the changes.

CUTTING IN OPTICALS

Although you recorded the opticals on your count sheets reel by reel in sequential order the dupe prints will be delivered to you in disorder and in sections over a period of a few days. There will be a number of opticals in each roll on cores tails out. Rewind them heads out in rolls or on reels, removing the cores as you rewind, and let your editor view the opticals in the moviola and approve them before breaking them down by individual optical number. When you have an exceptional number of opticals or when you receive a large number of opticals periodically during the editing process, you might want to code them, but this is not usually done on episodic TV.

The optical department scribes your assigned effects number ahead of each optical. As you separate each one into an individual roll, line them up on your bench shelf left to right in reel and effects number order. *Cutting in opticals* means replacing the original work prints with the optical prints or dupes. Do not wait for all your opticals to arrive. As soon as you can do so, cut in the first batch.

Begin with the optical roll at the extreme left on the bench shelf. This should be the lowest-numbered optical in reel 1 in our hypothetical case, the fade in 101 (see Figure 14–1). In addition to the optical number scribed in the extra footage at the head, your original key number (using only the last three digits) should also be scribed in the one-foot handle at the zero frame. Since you ordered this optical to be duped from B15X27942, it would probably be scribed *942* at the zero frame.

The full in of this three-foot fade in is at 946 + 02, which means that the first frame of the picture is 943 + 03. Place the film in the first gang of the synchronizer at the 944 + 0 frame. Set the framer at the zero frame. You need not take the time to set your counter at 944, but if you wish to do so, it will take only a second to turn it to 4444. Now, back up two feet. You are actually in the black leader, as prescribed by the network format 7 in Figure 13–3, but hypothetically you are at 942 + 0. Now place the optical print in the second gang at the scribed 942 in sync with your work print. Sometimes an optical company will voluntarily add another foot or two to your handle thus scribing an earlier key number, and you will have to back up accordingly.

Do not rely entirely on this for your sync. By *eyeballing* or by using a magnifying glass select and mark with your grease pencil at least two disparate frames of action on your work picture. Find the same frames on the dupe print and mark them; if the marks line up when you roll down to them, you can be assured the two prints are in sync. There is always some specific frame that is particularly identifiable. For example as a man walks down a street there is an exact frame where his right or left foot first touches a crack in the sidewalk, or as he passes a tree a frame where his right hand *leaves* the tree. Even in a closeup of a woman's face, there will be a slight enough movement of her head so that you can mark a frame where her left ear *touches* and then *leaves* the lamp in the background.

If the scene is such that you have to use a moviola to match action, it might be preferable to reverse the above procedure by first finding those matching frames. Place both prints in the synchronizer, line up your marks, and if all marks are in sync, back up to the scribed number on the dupe to verify its relationship to your original key number. If the key numbers are in sync, separate the splice where your first frame, 943 + 03, is joined to the end of the black leader. Cut the dupe at that exact frame line and splice it to the leader.

With the two prints in sync you must now check whether the optical itself is correct. Is the fade in complete at 946 + 02? Some optical houses will scribe two or three *x*s on the left side of the count frame that in this case is at the full in. They may also scribe another original key number reference either before or after the count. If you are satisfied that all these areas are in sync when you arrive at the end of the cut, mark the dupe even with the splice on the work print, separate the splice, cut the dupe at the mark, and splice it to the head of the following scene. Some optical companies do not voluntarily scribe in the key numbers and start out and full in marks. You have to request that they do so.

All of your opticals are cut in as described. In dissolves and other effects such as in effects 205 and 303, you should match action and recheck sync on both outgoing *A* scenes and incoming *B* scenes. Your sample count sheet illustrates only a few simple optical counts. Actually effect 101 would probably begin the supered titles over certain scenes for the opening credits.

You must understand that each optical represented by an effects number is *one complete unit* regardless of its length or the number of effects with only *two* splices required—one at the beginning and one at the end. The dupe print for optical 303, for example, will be spliced into your work print at the head of 303, the fade in, and at the end of 303D, a cut. Therefore it becomes obvious how vital it is that you be absolutely accurate when you order opticals because one slight error may require reordering an *entire* optical as long as or longer than 303.

What do you do if a scribed key number or full in or start out mark on a dupe does not sync with your original print even for only one frame? As a new assistant, you should *not* cut in the dupe until you have consulted with your editor. If mostly action scenes are involved, your editor can usually make the optical usable by making a slight adjustment. If most of the scenes are dialogue, it may be necessary to redo the optical. If the optical company made the error, a phone call should suffice, but if the error was yours, you may phone in the correction but you may also have to send a corrected count sheet.

OPTICAL TRIMS

The handles, or heads and tails, of the opticals are your *optical,* or *dupe, trims,* and the original prints being replaced are the *original trims.* Some assistants roll up their trims by reel numbers with dupe trims either separated from the original trims or all rolled up together and with any late opticals filed in the trim box for the proper reel as individual rolls.

My preference is for individual rolls by effect number with both optical and original trims together and with the trim boxes filed numerically by effect numbers. If each roll is clearly identified with its effect number, either the dupe or the original can be located most efficiently.

OPTICAL RECORDS

All your count sheets and duplicate order forms pertaining to opticals, main title, end credits, or format material such as bumpers, teasers, or trailers should be kept accessible. A good practice is to keep such material on a clipboard hanging from a hook or nail above your bench and to file it into a notebook only when the work has been completed.

As you cut in each optical indicate this on your count sheet. You usually receive most of the material you ordered within a two-day period. The remainder will come in piecemeal so it is important that you maintain some sort of record of missing items to avoid having to leaf through a lot of orders and count sheets. You must stay on top of everything. If too much time elapses, contact the optical department to learn when you might expect any missing opticals. Should the post-production supervisor call regarding the progress of the completed picture, you will be able to give her immediate answers.

The *completion record* (Figure 16–1) is a suggested form that you might use for recording the receipt of the piecemeal opticals after the bulk of the effects have been received. In the *material missing* column note anything that is missing from each reel—opticals by effect number, format material such as titles or credits, or an insert or pick up still to be shot. As you receive each one, line it out and indicate the date you cut it into the picture. Also record the dates each reel is keyed, has its negative cutting completed, and is dubbed.

	MATERIAL MISSING:	DATE KEYED	DATE SENT INCOMPLETE	DATE SENT COMPLETED
REEL 1—	~~TEASER~~ 11/2 ~~MAIN TITLE~~ 11/3 ~~#102~~ 11/3 ~~(OPENING CREDITS)~~	10/26	10/31	11/3
REEL 2—		10/26		10/31
REEL 3—	~~#303~~ 11/1	10/27	10/31	11/1
REEL 4—	~~#402~~ 11/1 ~~#403~~ 11/2 ~~#405~~ 11/2	10/27	10/31	11/2
REEL 5—	~~#501~~ 11/1 ~~#503~~ 11/2 ~~END CREDITS~~ 11/3	10/27	10/31	11/3

DATE DUBBED: 11/2, 11/3 DATE/ANSWER PRINTED: 11/7

Figure 16–1. Assistant's completion record.

NEGATIVE CUTTING

The negative cutter matches each cut in the edited picture by key number with corresponding negative. She will match internegs to your dupe prints, which have *dupe key numbers.* The former dupe number, 37X42369, was easily distinguished from an original key that began with a letter. You must remember, however, that the more recent dupe key numbers begin with *A* or *D,* i.e., A25X29158 or D15X79360.

Keying

In a situation where there is even less than ordinary time the post-production supervisor may instruct you to send all the reels to negative cutting as soon as you have turned the picture over to M&E and completed all optical counts and orders. Tape together all the picture reels and address them with the notation that they are *for keying only.* The negative cutter must *not* cut any negative but will only *log* the key numbers except those comprising optical effects, which she will be able to identify by the lines drawn on the film. The latter key numbers will eventually be identified by dupe key numbers for the internegs. The negative cutter should return the reels to you as soon as possible, but the log will enable her to pull out the negatives needed and have them lined up on the bench by the time you return the reels for cutting. This will then save her considerable time.

Incompleted Reels

Ideally, you retain each reel until it is complete, meaning that all effects and credits or titles, if any, have been cut in. Then you send it to the negative cutter, marking it as *completed.* However, you usually can not afford the time for this with impending television air dates. If one or more of your reels have only a couple of effects missing and you have been advised that you will not receive one of them for another day or so, valuable time can be saved by sending the reel, marked *incomplete,* to the negative cutter.

First wind down in the reel to the missing effect and, ahead of the head splice, in the center of the last few frames of the preceding cut, place a tape inscribed with the warning, ◄—— *DO NOT CUT NEGATIVE FROM HERE.* After the end splice of the complete optical over the first few frames of the next cut, place a second warning, *DO NOT CUT NEGATIVE TO HERE* ——►. The negative cutter will not match negative to anything between those two tapes. She will cut negative on everything else in the reel and immediately return the reel to you for completion.

FINAL RECORDS

Your objective is to have all your opticals and titles cut in by dubbing so that M&E editors and the mixers on the dubbing stage who work with only B&W reversals can review the final mix of each reel in color. It will also be screened for the producer and will give her the opportunity of approving the effects and credits. Your completion record will hopefully be completed by then. If, however, the negative cutter is still working with some of the reels, each must be sent for and returned to her after the screenings.

Your *assistant's screening record* (see Figure 12–2) shows your footages for the first two screenings. Using the same form Figure 16–2 lists the final cut and final footages. If a *final footage record* (Figure 16–3) is requested by your post-production supervisor, it is not necessary to remeasure the reels. This form requests information relating to the network format (see Figure 13–3) that you can provide by referring to the format and to the final footages assuming, of course, that the garbage (format material) you inserted in the picture was accurately measured.

Beginning with the first column from the left, the final footage record asks for *final footage including all format material.* Since your personal record of the final footage lists *only action footage,* you must add the format footages indicated in Figure 13–3. For example your footage for reel 1 is 781 feet + 11. At the head of reel 1 is format material 1 through 7 totaling 146 feet. 781 feet + 11 plus 146 feet equals 927 feet + 11. Reel 2 is 841 + 02; the garbage between acts I and II (9 through 11) totals 7 feet + 0; 841 + 02 plus 7 feet equals *848 feet + 02.* The remaining reels in the first column are estimated in the same manner.

The second column asks for *each* commercial banner location *from first frame in reel.* You must still include format footages. If a reel does not have a commercial banner, it has to be left blank. If you have a reel containing more than one banner, you have to list each banner footage in the reel. Note that the numbers of all the commercial banners and the one station break are specially indicated by parentheses on the format.

The only commercial banner in reel 1, (6), is at the head of the reel before act I begins. When we add all the format material preceding the commercial, 1 through 5, we conclude that the banner is *142 feet + 0* from the first frame in the reel. The

DATE FOR WHOM	BEG. SC. –	DESCRIPTION	REEL	REEL FTGS	BREAK FTGS	ACT	ACT FTGS
10/25		Estab. San Diego, day	1	781+11	781+11	I	1028 + 05
Final Cut	14	Detectives to Kitchen	2	841+02	246+10 594+08	II	1093+12
Ned Forbes	34D	Marshall in apt: "Am I under arrest?"	3	823+08	499+04 324+04	III	1049+0
	46A	2/5 Ryan/Kirkwood in car	4	941+13	724+12 217+01	IV	1023+07
	65F	Kirkwood: "Ryan will take care of that."	5	806+06	806+06		
		TOTALS—		4194+08	4194+08		4194 + 08
		Ftg. Over:		7 + 08		Time Over:	5 secs.
		Ftg. Under:				Time Under:	
10/25			1	781+11	781+11	I	1028+05
Final Footage			2	841+02	246+10 594+08	II	1093+12
			3	823+08	499+04 324+04	III	1049+0
			4	934+05	724+12 209+09	IV	1015+15
			5	806+06	806+06		
		TOTALS—		4187+0	4187+0		4187+0
		Ftg. Over:				Time Over:	
		Ftg. Under:		none		Time Under:	none

Figure 16–2. Assistant's screening record.

commercial banner (10) in reel 2 is preceded by 246 feet + 10 of action footage (from the final footage in Figure 16–2) plus 3 feet of black, 9, equals *249 feet + 10.* Reels 3 and 4 are similarly calculated. Reel 5 has *two* commercials, (27) and (30), in the end format material. When you add the action footage, 806 feet + 06, and 3 feet of black, 26, the first banner is at *809 feet + 06* and the second banner when you add from 809 feet + 06, is at *813 feet + 06.*

The last column of the final footage record requests only the *station break* location, or footage, from the first frame of the first reel, still including format material. The station break (17) is between acts II and III in reel 3. In the first column the total footage of the first two reels is 1775 feet + 13. In the middle column the footage to the commercial banner in reel 3 is 502 feet + 04. For the commercial (14) itself through the first bumper, 16, to the station break there is an additional 13 feet + 08. Add 1775 + 13 plus 502 + 04 plus 13 + 08. The station break is *2291 feet + 09* from the first frame in reel 1.

This report can be used by the post-production supervisor to verify that your episode complies with the *Note B* requisition of the network format (see Figure 13–3

COLUMBIA PICTURES TELEVISION
FINAL FOOTAGE RECORD

PROD.# 8191 TITLE: Murder, Anyone?
PRODUCER: Forbes DIRECTOR: Jason
EDITOR: Bernard ASSISTANT: Rose
DATE LAST OPT. ORDERED: 10/26 DATE LAST TITLE ORDERED: 10/26
DATE LAST INSERT SHOT: 10/21 DATE SHOW TURNED OVER: 10/26

THIS FORM IS TO BE COMPLETED AND DELIVERED TO THE SUPERVISING ASSISTANT ON THE SAME DATE THE LAST REEL OF THIS PRODUCTION IS TURNED OVER TO MUSIC AND SOUND EFFECTS.

FINAL FOOTAGE INCLUDING ALL FORMAT MATERIAL		COMM'L BANNER LOCATION FROM 1ST FRAME IN REEL		STATION BREAK LOCATION FROM 1ST FRAM OF RL #1	
REEL 1	927 + 11	REEL 1	142 + 0	REEL 1	
REEL 2	848 + 02	REEL 2	249 + 10	REEL 2	
REEL 3	853 + 08	REEL 3	502 + 04	REEL 3	2291 + 09
REEL 4	941 + 05	REEL 4	727 + 12	REEL 4	
REEL 5	862 + 06	REEL 5	809 + 06 813 + 06	REEL 5	
REEL 6		REEL		REEL 6	
REEL 7		REEL		REEL 7	
REEL 8		REEL		REEL 8	
REEL 9		REEL		REEL 9	
REEL 10		REEL		REEL 10	
REEL 11		REEL		REEL 11	
REEL 12		REEL		REEL 12	
TOTAL DUB. FOOTAGE	4433	FORMAT DUB. FOOTAGE	246	OVER UNDER	none

NOTE: IF THE STATION BREAK DOES NOT FALL WITHIN THE LIMITS PRESCRIBED ON THE FORMAT, CALL THE POST-PRODUCTION SUPERVISOR .. IMMEDIATELY.

IF ANY OF THE ABOVE FOOTAGES SHOULD CHANGE AS A RESULT OF POST-FINAL EDITING, SUBMIT A REVISED FINAL FOOTAGE RECORD TO THE SUPERVISING ASSISTANT .. IMMEDIATELY.

Figure 16–3. Final footage record.

page 2). It requires first that the total *action footage* of acts I and II *must fall between 1810 and 2710 feet.* Is this the case? The total footage for the two acts, as recorded in Figure 16–2, is *2122 feet + 01.* The second requirement in *Note B* is that the station break *must fall between 1980 and 2880 feet, including format material.* Based on the final footage record, the episode also fulfills this requirement.

DUBBING

The dubbing stage has an electronic panel controlling the mix of all the sound elements that will become a vital counterbalance with the visual film. There are usually three *mixers* (in Hollywood these would be members of IATSE Sound Technicians Local 695) operating the panel. The head mixer regulates the dialogue, and the other two technicians control the sound effects and music.

A melange of hidden wires connects the buttons and levers of the panel with a *dummy,* or *machine, room* that contains rows of huge sound machines running all the units built by M&E. A projectionist in a booth high in the rear of the dubbing stage runs the picture in sync with all the units for each reel.

The sound editor responsible for each reel is on the stage as her reel is being mixed. The exigencies of on-going episodes do not allow the producer the time to supervise the dubbing. Instead, she usually assigns this to an associate producer. The producer views the final dub and either approves it or requests that certain areas be *sweetened,* meaning that the effects should be increased or diminished.

By this time you and your editor are too occupied with the next episode to be involved in the dubbing unless the editor is called regarding some problem. The music editor who is also working on another show might not be able to attend but will be *on call.*

Your principal responsibility is to make certain each reel of your work print (W/P) is delivered to the dubbing projection booth in time for its final check and run through. If the negative cutter is still working on your reels, you may have to get transportation to bicycle them back and forth.

Dubbing sessions on most episodes are usually scheduled for completion in one day in order to ensure the air date and minimize dubbing expense. Editorial changes, errors by M&E editors, or a lot of sweetening jeopardize this schedule. When some of these incidents occur, it often results in extensive and expensive overtime. Dubbing sessions cannot often be extended into the next day because the stage is usually booked for another project. Your completion record (see Figure 16–1), however, indicates that dubbing on your hypothetical show did extend into a second day.

Most studios have several dubbing stages, and all are kept constantly booked during the television season. It is not advisable to shift a picture in the middle of dubbing from one stage to another with different mixers. Familiarity with a project is an asset in dubbing. Therefore one trio of mixers generally do all the episodes of a series.

Should the associate producer request an effect that was not anticipated, the sound editor would have to rush back to her cutting room, get the effect from the sound library, possibly get a transfer from the sound department, and then edit the effect into a unit. All this precious time is wasted as the mixers and others wait on the dubbing stage unless it is practical to go on to another reel. Sometimes the delay can be avoided if the effect can be found in the dummy room where a limited library of ¼-inch sound effects tapes are kept for such emergencies.

COMPOSITE ANSWER PRINT

The final dubbed sound of all the reels is a three-stripe on full coat stock with *A, B,* and *C* channels for dialogue, music, and sound effects respectively. From this a composite, or three-to-one, single-stripe full coat master and an *optical track* are made. The optical track is photographic sound in the form of a black band of visible striations, a type of sound recording used from the inception of *talkies* until it was replaced by magnetic track.

Using the head pops, which you placed at the three-foot mark in the academy leaders, and the tail pop inserted by the sound editors usually three feet past the last frame of each reel, the negative cutter syncs the optical tracks to the cut-negative picture reels. The lab makes a *composite print* by printing an optical track on the left side of the picture aperture thereby combining sound and picture on a single film.

The first, or *answer, print* is sent to your studio for approval. It is screened for either the producer, the associate producer, or the post-production supervisor who may request color or density corrections in certain scenes that would require retiming by the lab. These changes must be absolutely necessary and should be kept at a minimum due to the limited time before the airing.

The post-production supervisor orders whatever is required by the network in terms of videotapes, release prints, and material for possible foreign distribution.

EXCEPTIONS

The longer post-production editing period devoted to features and television movies as compared to episodic TV, results in more care and attention being given in an effort to make the project as good as possible. There is more time for experimentation and refining. The producer and possibly the director, depending upon the director's contractual agreement, personally follow through the finalizing of the picture by supervising the ADR, scoring, and dubbing sessions.

The film editor is able to join them because there are no overlapping episodes to divert her efforts. The assistant editor can also attend unless she has to remain in the cutting room to cover the phone in case of any important calls. Sometimes calls can be relayed by the studio operator to where the editor is, or there may be a second assistant or an apprentice on the project who can remain in the cutting room.

With larger budgets to work with in features and larger TV projects more time is permitted for everything. Only one or two reels may be dubbed in one day as compared to the five reels on your series dubbed in that same day. There is more time for trial and error, additions, and deletions.

Editing changes are sometimes made even during dubbing after negative has already been cut. Your editor is then under tremendous pressure to use all her editing and creative ability to make such changes quickly and skillfully. She will need your help to find trims and as a last resort may instruct you to rush through an order for a print from *B* negative to circumvent negative cuts. If the project is a television movie, overall footage has to be maintained as on a series episode, but if it is a feature, footage is not a major concern. However, the editor will try to minimize the amount of changes that must be made in M&E dupes and units.

On projects produced by an independent company that does not usually employ a post-production supervisor, a *production manager* is usually hired preproduction to arrange for locations, crews, and lab and editorial facilities. She manages the budget and schedules dubbing. In many cases, however, she is dismissed after shooting and the producer takes over her post-production responsibilities. This often results in the editor being instructed to carry out some of these duties, which in turn are passed on to the assistant.

CHAPTER 17

IT'S A WRAP!

Throughout the preceding pages you have experienced becoming an apprentice film editor in a major Hollywood film company or studio then being promoted and serving as an assistant editor on a one-hour television series.

The evolution of the first episode of your TV series has been described in detail. The progression in the rest of the episodes or on a long-form television project or feature is basically the same. If you now understand clearly who does what, when, and why, you have gained a sense of the making of a film from an editorial and post-production perspective. You should now know the separation of duties between the apprentice and assistant editors and the film editor, and you should understand the sequence of work from dailies through first cut to the final cut and the turning over of the picture to music and effects, the ordering and cutting in of opticals, and finally negative cutting, dubbing, and the answer print.

WRAPPING THE PICTURE

After delivery of the answer print, you can rest assured that the first episode is complete and all material relating to it in the cutting room can be removed to make more room for the second episode, which by this time you and your editor are already editing. Remove the earlier episode's trim boxes from the racks and pile them in a corner or in the hall. Place all papers—code sheets, daily forms orders, lined script, and so forth—into a large manila envelope identified with series name and episode production number. Notify shipping and receiving that the paper work and trims are ready to be picked up and properly filed.

ASSISTANT/EDITOR RELATIONSHIP

That which begins as only an occupational association between you and your editor will probably become a relationship not unlike a marriage. It is certainly dissimilar from the 9:00 to 5:00 employee/employer relationship common in most other businesses. The exigencies of movie-making are responsible for the difference.

You have been warned that the demands on your time are often exorbitant. It is not an exaggeration to predict that you and your editor may spend more time with each other than you do with your spouses, who have to understand that this is an unfortunate prerequisite of your careers.

An assistant must be many persons for his editor. He must *manage* the cutting room so the editor can concentrate as much as possible on editing with minimum distractions. He must be extremely well organized so that any paper or film, whether it be a small delivery slip or only one frame or one sprocket of picture, can be found immediately. He must be efficient and accurate while also being reasonably fast particularly when he is under tremendous pressure. Even more important he must be consistently conscientious and reliable.

Whenever you have to leave the cutting room, let the editor know where you are going and when you will be back. Return promptly, but if you are unavoidably delayed, phone him so that he does not wonder what happened to you. Should you be left in charge of the cutting room while your editor is on the dubbing stage or elsewhere, do not use this occasion as an opportunity to disappear. You will be held responsible for any important messages that are missed. Protect your editor and the project at all times.

Here are a few more *do nots*. Do not become habitually tardy or negligent, requiring you or your editor to request additional help.

Do not speak out of turn to your director or producer, regardless of how friendly they are. If you have a suggestion that you are convinced will improve the project, first speak privately to your editor and let him decide whether it should go any further, and whether you or he should present it.

Do not ridicule the project you are working on, even though it may be the worst film since *The Attack of the 50-Foot Woman.* Your editor is aware of it and does not want to be reminded. Instead, he is concentrating on trying his best to make the show as good as possible. Try to be encouraging and optimistic. It will certainly help your editor, and you too, get through an unrewarding project as pleasantly as possible.

Do not divulge personal comments made in the cutting room or projection room. Producers and directors assume that you will respect confidences, and you may become your editor's confidant by lending a sympathetic ear to his and his family's most intimate problems. Acquiring empathy for him and trying to understand him better will help you to anticipate his editing needs as well as his moods.

You extend so many diverse services for your editor, that it is admittedly somewhat one sided. What then does your editor do for you? It is regrettable that there are a few editors who do not reciprocate in any manner. They are not interested in your problems or in you as a person. They take but do not give. They may even relieve themselves of some responsibilities by taking advantage of your diligence and ability. Do not be disheartened by this lack of appreciation.

Your objective should be to become the very best assistant and not just an *okay* or *good* assistant. Know that you have done the best you could, that best being better than any other assistant or at least as fine. That feeling of self-satisfaction is reward in itself. Finally being a top notch assistant will not go unnoticed nor unrewarded in the long run.

Fortunately, most editors are very appreciative of worthy assistants. Your editor will reciprocate by acts of consideration and, if he is able, will help you get employment and give you sequences to cut as you learn to become an editor. When the opportunity arises if your editor believes you are ready to be an editor, he will be happy to recommend you.

SURVIVING

Survival in motion pictures is simply working as continuously as possible. The easiest way for an assistant to accomplish this is either to remain with a studio that maintains considerable production or to align himself with an editor who is consistently employed. Working on a television series as you have been (hypothetically, of course), your alliance with your editor will depend on the longevity of the series and the status of your company.

Four major companies—Columbia, Disney, Fox, and Warner Brothers—still have assistants who retain studio lot seniority. Such seniority can no longer be earned. If you are at one of the above studios, as a new assistant you are low man on the totem pole. Even if your series is a winner if other series are canceled and there is not enough work for the senior assistants, you could be *bumped.* Being a staff assistant for a successful company, however, may keep you working regardless of whether your series is renewed.

It is not usually possible to continue with a series editor as he moves on to assignments at other studios. Series editors have to take assistants assigned to them. Editors on long-form television and features generally have their choice of assistants, so this type of editor you might accompany to assignments at other companies.

Although staying at one studio may give you a sense of security and of being comfortably part of a large *family,* your job is predicated on volatile series ratings. A bad season for your company may force you to look for a job elsewhere, and if you have been at one place for several years, you may have few contacts elsewhere.

Being a topnotch assistant will be your asset. Editors with whom you have worked will vouch for you and try to help you. Other assistants who have passed through your studio will know of your reputation and if you call them will try to help you find an assignment. Word of mouth travels quickly in the industry. There are many run-of-the-mill assistants and not many who are first rate, and you may find that you are in demand and have to choose between several different assignments.

Whichever kind of assistant you are, the job will rarely come to you. You will have to find the job. In accomplishing this you should be as organized as you are at work in a cutting room. Remember what you had to do to become a member of this industry—checking the trade papers for forthcoming productions and communicating with all-important contacts. You must not limit that communication only to when you need help. It is a two-way street. Keep in touch continually. Perhaps you will be able to help others when they are looking for jobs. Obviously, *hitching your wagon* to a successful feature or TV-movie editor will result in more guaranteed employment. It will also offer you experience on films that provide more time for editorial experimentation and refinement.

Do not reject an opportunity to work in M&E, for a commercial house, or on a documentary just because your aspiration is to become a film editor. They are excellent backgrounds for editing and becoming more familiar with these other areas may result in your making a better evaluation of your career objective.

BECOMING AN EDITOR

How do you become an editor? You learn by observation, asking questions, and finally actually editing. While you are an assistant you can do these only when you have completed all your assistant's work and can afford the time to, with your editor's approval, watch him as he edits. Some editors cannot edit with someone looking over their shoulders; others do not mind as long as you do not talk; and still others can carry on an instructive dialogue while they edit.

After you have viewed dailies, consider how you would edit a particular scene. Then when your editor has edited, compare your *cut* with his. After screenings, imagine how you would make certain changes that were requested by the director or producer and compare your *changes* with those finally made by your editor. Most editors will answer your questions regarding why they edited a certain scene as they did but be diplomatic about when you question them. Try not to interrupt them for any reason when they are concentrating on a difficult scene or when they are under pressure.

Merely observing, listening, and asking questions, however, does not make one an editor. One must *edit*. You can practice editing by working on discarded trims from a completed show. You will probably have to do it on your own time, and it will not be easy to find that time after you have put in long hours. But if you are determined, you will find the time and the energy. If you have a good relationship with your editor and you trust and respect him, ask him to criticize your work. Then make the changes your editor has suggested and evaluate the reasons for them by the improvement in the scene.

Your next step is to ask your editor if you might cut a sequence for him, but do not act hastily. First, be certain that your work as an assistant is absolutely irreproachable. Otherwise why should your editor respect you enough to let you cut his film? Are you really ready? If you are deluding yourself, you may make a mess of that sequence and cause your editor more work than he deserves. And should that happen, what do you think your chances are of getting another sequence to cut? Finally choose the right time to ask, a time convenient for the editor. These questions are not asked to scare you or discourage you, but to help you make the right decision. When you do begin cutting sequences, never let it interfere with your assistant's work. That has first priority.

Some editors may not let you cut a sequence regardless of how much they respect your work as an assistant. Do not think harshly of them or take it personally. There are editors who are so possessive about their film they would not want anyone else to cut any of it, even a fellow-editor. Your editor may feel insecure or nervous about the particular project or be under extreme time pressure, and even though he could use some editing assistance, your help may cause him additional work, and he cannot take the risk at this time.

Generally, however, once you have earned his respect for your editing your editor will not hesitate to let you cut sequences when it is possible. In fact beware of the few editors who, in order to give themselves free time, take advantage of you by letting you do an exorbitant amount of the editing and who also neglect to mention your contribution to the director or producer. You may be willing to accept this imposition, however, in return for the invaluable experience. Most editors will go out of their way to credit a good editing job by their assistants thereby posing them for consideration in the event of an available editing assignment.

You must have been a member of the Guild for five years before you can be assigned as an editor. During this time get as much practice editing as you can so that you will be ready when an editing assignment is offered. Episodic television provides the most opportunities for such a promotion particularly if you have been with a company for a considerable time.

A promotion may take longer for the assistant working on features or long-form television, but if you have been an editor's loyal assistant through many projects, he may finally be able to make you a coeditor on one of his films. With this kind of screen credit you may be able to continue editing such projects, although it will be much more difficult getting jobs than it would be if you were a series editor.

At first you may *bounce* back and forth, working as both assistant and editor as the assignments become available to you, but at some point you must make a choice and decide whether your career is to be assisting or editing. If you do not make this decision, eventually you will not be taken seriously as an editor and yet employers will hesitate to offer you an assistant's position. Your choice will be predicated on your confidence in yourself and on your economic situation.

The economic factor is determined by the instability of the motion picture industry, its seasonal unemployment, and the increased competition of your fellow Guild members. There are more editors than there are assistants competing for the available editing jobs even when the industry is at its busiest. As more and more aspiring editors complete their mandatory five years of eligibility, this inequality increases. You may overcome this by remembering those three Ps—passion, patience, and perseverance.

REVIEWING THE VIDEOTAPE VIEW

Three innovative electronic systems, the Ediflex, the EditDroid, and the Montage have been principally responsible for the significant penetration of videotape into Hollywood's editing fraternity during the past few years. These systems succeeded mainly by providing the means of making an editing change in a scene without being forced to redo or duplicate (dupe) the rest of the tape as was previously required by other videotape equipment. This, combined with the option of shooting on film, transferring to tape for editing, and either remaining with tape for final distribution and viewing or by matching time codes in negative cutting returning to film has made our beloved moviolas and flatbeds endangered species.

Motion picture editor Elio Zarmati wrote in the Guild's *Newsletter* (November, 1985)

> . . . I think that, in a few years, we'll all be using Ediflexes and Montages and Edit-Droids or whatever else some whiz kids are concocting in their garages, attics and research labs. . . . Electronic editing is cleaner, faster and light years ahead of the green machines still regarded as the standard tool by more film and television editors.

Most of the recent videotape systems have used film editors as consultants in the development of their equipment and continue to make adjustments and improvements based on their suggestions. The machines have been designed specifically for those experienced in editing film because merchandisers know skeptical producers and directors are more apt to try their products with film editors they are used to relying upon.

It will be interesting to see how many production companies gradually succumb to the wizardry of electronics in the next few years. Technological progress is always slow at first until the *wrinkles* are ironed out and the price for equipment becomes less. Editing systems are still costly to purchase and expensive to rent. They may be ". . . cleaner, faster and light-years ahead. . . ." but only one editor can work on the machine at a time and editorial decisions by directors and producers take the same amount of time whether the picture is on tape or film. Since the use of videotape can be extended in post-production to sound effects and the use of synthesized music, the cost factor is mostly resolved by the decrease in necessary personnel. Whereas editing a one-hour show on film required three editors and three assistants, using the new electronic systems has reduced that number in some cases to two editors, one assistant, and one apprentice. Since other reductions in personnel also occur in other areas of post-production, the effect on employment is obvious.

One can not dispute the attraction of videotape. There are no *foot cuts* nor is there aggravating film to unravel, and no footage should be lost. A first cut, or any other screening, can be easily saved and reshown at any time. Simple opticals accomplishable on some off-line machines and others including very complicated effects performed with such awesome immediacy on line offer tremendous benefits. Finally aside from any delay because of editorial decisions, there is no doubt that videotape editing is much faster than is film editing.

As equipment costs decrease, more television productions will undoubtedly climb aboard the videotape band wagon. We can only hope that that economic factor, combined with continued technical improvements, will result in more production especially for home video cassettes and cable programming and so will provide additional jobs.

It behooves anyone interested in any kind of editing career to be aware that as technology continues to make remarkable advances tape will play an increasingly important role in the production of motion pictures. Will this make obsolete film and all the film equipment such as moviolas and flatbeds, which occupy hundreds of editing rooms and are still being sold? I do not believe so. Film and tape will probably harmoniously cohabit the industry as they already do for some time. So long as film remains, you owe it to yourself to get your editing experience in *film* to better prepare yourself for the future.

IN CONCLUSION

I have painted a rather bleak picture: the difficulty of securing a union job; the habitual long hours, pressure, and family sacrifice; and the constant threat of unemployment. I never said it was going to be easy. You really need those three Ps. If you have them and really harbor the dream and passion to devote your life to making movies, you must realize there is a reverse side to that bleak picture.

For example those long, arduous days when you had to eat your lunches *on the run* are sometimes relieved by some days of relative tranquility when a considerate editor will let his assistant take some long sociable lunches or perform some personal errands. There are the morning breakfasts with your fellow workers, the sometimes elegant and always enjoyable wrap parties or Xmas celebrations with cast and crew, and the company screenings and picnics or athletic events. There are additional screenings and seminars offered by the Guild.

There is that wonderful excitement of receiving your first screen credit as an assistant or apprentice, probably on a long-form TV or feature. Such credits are rarely ever given on an episodic series.

There are the wonderful fringe benefits of vacation and holiday pay, retirement, and excellent medical insurance. There are the good salaries. Enjoy but always try to save enough for those *rainy* periods of unemployment. At all studios and most large companies a credit union will be available to offer you a convenient method of forced savings.

There is the continual wonderment and anticipation about what the next project will be and what it will bring. There is the stimulating hope as you strive for promotion, the excitement of first becoming an assistant, the exhilaration and pride of becoming an editor, and the thrill of that first editing credit on the screen.

Now as an editor it begins all over again—the aspiration and struggle for worthwhile assignments; the acceptance for membership in the American Cinema Editors, the Television Academy, or the Motion Picture Academy; and the hope of being nominated for, and perhaps winning, an Eddie, an Emmy, or an Oscar.

Finally, there is that wonderful feeling of satisfaction that you have been selected to work alongside talented, creative people making a movie that will be viewed by millions of viewers. Whether you are an apprentice or an editor you should anticipate going to work each day and enjoy being there. How many individuals that you know can honestly say that about their jobs?

I can truthfully say that I did and still do. I hope you will feel the same and trust that this text will have helped you.

GLOSSARY

A&B Rolls Alternate negative cutting on two (A&B) rolls. With the subsequent A&B printing process some opticals and titles can be effected thereby eliminating the necessity of preordering them.

Academy leader Beginning and end leaders used on work picture, work track, and release prints. The head leaders have a 12-foot start mark.

Acmade or **Acmade Codemaster** A light, portable single-reel British coding machine that uses various colored tapes to *emboss the numbers on the film.* (*See Acmade coding*)

Acmade coding Code numbers on an Acmade machine consist of either seven or eight characters. They begin with three numerals and end with four numerals. In between those two sets of numerals can be a blank, a 1, or one of the following letters: A, B, C, D, E, M, W, or P. Sample code numbers would be 038 B 0000, 514 P 9361, and 000 3000.

Action footage Measurement of only the story portion of a picture.

Adaptor A small plastic core that fits onto a male flange converting it to a female flange for the rolling up of film. (*See Flange*)

ADR, automated dialogue rerecording or **automatic dialogue replacement** The use of an electronic system to arrive at sections of a cut picture that require dialogue be redone by the performer in order to replace the production track thus improving both sound and performance quality. Still often loosely referred to as *looping.* (*See Loop*)

Ambience The sound of the background atmosphere. (*See Room tone*)

American Cinema Editors (ACE) Hollywood-based honorary society of film and videotape editors identified by "A.C.E." following their names on screen credits.

American or **standard coding** Five or six characters consisting of one or two letters (A through K except I) and four numerals. Examples are A 1000, DB 9371, KK 0000, and so forth. (*See Numerical slate coding*)

American coding machine A large, heavy, self-standing machine that has been standard coding equipment for over 50 years beginning with the single-reel system that codes one reel at a time and more recently with the double-reel system on which picture and track can be coded simultaneously. (*See American coding*)

Answer print The first composite print of the completed film, generally used to approve the color and density of each scene so that any necessary corrections might be made before release printing. (*See Composite*)

Aperture The opening in a camera that provides light to film a scene or as in a moviola or projector to view a film. In projection TV aperture requires a 1.33 lens, standard feature aperture is 1.85, and wide screen aperture for Cinemascope, Panovision, and so forth is 2.35.

***A* scene** or ***A* side** The right side of the cut; the scene being cut *from;* the outgoing scene. (*See B scene*)

Assembly The splicing together of a number of takes or scenes as with dailies. Lifting out certain scenes from the dailies for a screening can also be referred to as an assembly since it does not require any editing. It is inappropriate, however, to refer to a *first cut* as a *first assembly.* (*See Cut*)

Backing Refers to that which is *behind* a positive print. Is there original negative? Interpositive? Or an internegative?

Balancing band The narrow stripe on a single-stripe magnetic track that is on the left side when the sound is heads up, mag up. It has no sound and serves only to maintain level contact with the sound head so that the best quality sound is produced. The wide stripe on the right side records the sound.

Band (*See Channel*)

Banners Short lengths of temporary film used during editing to identify forthcoming material such as *scene missing, insert missing, stock missing, commercial,* and *freeze frame.*

Base (*See Cell*)

Basic show A TV term. (*See Action footage*)

Beep (*See Pop*)

Birds eyes (*See Changeover cues*)

Black & white (B&W) reversal A print made directly from a black and white print or color print. It is primarily used as a duplicate work picture for music and sound effects editors. Also called a *B&W print, reversal print, dupe print,* or *dirty dupe.*

Black frame leader Black opaque film with visible frame lines usually used as a temporary substitute for missing frames of picture.

Block (*noun*) Used in reference to a synchronizer as a sync block or to the numbering apparatus on an American coding machine as the numbering block. (*verb*) (*See Block in* and *Block out*)

Block in or **Close in** To verify the exact footage of an optical by completing the editor's markings at the full in and start out.

Block out The superimposition of dissimilar numbers over erroneous code numbers to make them illegible so that the film may be recoded correctly.

Blowup The enlargement of a cut, a scene, or an entire film by optical printing. (*See Move-in*)

Blowup chart A chart on film provided by an optical facility that enables an editor to select the size of a blowup by number. (*See Blowup*)

***B* Negative** The negative of all the takes of a film that the director has not ordered printed but that are developed and carefully filed.

Break A television idiom for an interruption of a film, such as an act break, a commercial break, and a station break.

Break down (*verb*) Separating certain track and picture takes for numerical sequence in the preparation of dailies; separating dailies into individual takes for the editor. (*noun*) The preproduction shooting schedule prepared by the assistant director.

Build (a) Creating a scene from beginning to end. (b) Progressing to a particularly important story point. (c) The preparation of format material for insertion into a television film by the assistant editor. (d) Editing various sound effects in sync with the work picture. (*See Units*)

Butt splicer A small device that has attachments to cut the film and to tape two pieces of film together. (*See Hot Splicer*)

***B* scene** or ***B* side** The left side of the cut; the scene being cut *to;* the incoming scene. (*See A scene*)

Camera report A form filled out by an assistant cameraman that lists all the takes of every scene for each day's shooting. The takes selected by the director for printing are circled on the report and irregularities, instructions, or other information pertinent to the lab are noted.

Cell The *base,* or *shiny,* side of the film as opposed to the *dull* side. (*See Emulsion*)

Changeover cues, changeovers, or **end cues** White grease pencil marks inscribed by the assistant editor at the end of each picture reel, except of course the last reel, to guide the projectionist in making a smooth changeover to the next reel. On composite prints the lab provides small light circles called *birds eyes.*

Channel The *band, stem,* or *stripe* on which sound is recorded on the track.

Circled take (*See Take*)

Clamp or **spring clamp** A metal device with a spring to hold two or more reels together on a rewind so they can be wound in unison.

Clapsticks A pair of hinged boards containing identification of the take, production, company, director, cameraman, date, and so forth. As the two boards are banged together, usually before the action of a take begins, the first split moment of impact provides a visual and audio sync between picture and track. Also called *markers, slates,* or *sticks.* (*See End markers*)

Code numbers The consecutive numbering of each take of picture and track at one-foot intervals for the purpose of synchronization and identification. (*See also American standard coding, Acmade coding, numerical slate coding,* and *Key numbers*)

Coding (*See Code numbers*)
Color work print (*See Work picture*)
Composite print Picture with dubbed dialogue, music, and sound effects on a single film. (*See Answer print* and *Optical track*)
Constant speed The left pedal of the moviola runs both picture and track at standard speed—90 feet per minute. When the *interlock* is released, only the track is operated by the constant pedal. (*See Variable speed*)
Core A small plastic implement onto which film may be wound when the core is attached to a *flange* or *split reel.*
Cueing or **cues** (*See Changeover cues, music cues,* and *spotting session*)
Cut (*noun*) The point at which an angle has been joined or spliced to another angle; the angle or piece of action preceding or following a splice; the screening of the edited film as in *director's cut, first cut,* or *final cut.* (*verb*) (*See Edit*)
Cut picture The edited film not necessarily finalized but in any edited stage.
Cutter An often-used misnomer for an *editor. (See Negative cutter*)
Cut trims Refers to material that was edited into the picture and later taken out.

Dailies The picture and track takes of each day's shooting that have been selected by the director for printing; the screening of these synchronized takes. Also called **rushes.**
Degauss or **demagnetize** To eliminate any undesirable particles from editorial equipment that could adversely effect the work track; to erase all sound from a magnetic track so that it can be reused.
Degausser An electrical device used to demagnitize editing equipment or magnetic track.
Dirty dupe (*See Black & white reversal*)
Dissolve An optical effect permitting a scene to fade out while simultaneously overlapping the fading in of the incoming scene.
Double printing or **multiple printing** Optically slowing down the action. (*See Overcrank*)
Drag A tension bolt sometimes inserted through the top of the left rewind to control the rotation of the shaft.
Dropout The disappearance or nonrecording of any sound in the track.
Dub (*noun*) A session on a *dubbing* stage during which two or more tracks are rerecorded or *mixed* into one composite track; usually refers to the final editing phase in which music, sound effects, and dialogue are mixed into one track. (*verb*) To rerecord.
Dupe (*noun*) A print made from a dupe negative or internegative; a B&W reversal print; a duplicate track. (*verb*) The process of ordering a dupe print or reversal print.
Dupe negative Second or lower generation B&W negative derived from a *fine grain.* Also loosely uscd in reference to an *internegative.*
Dupe print (*See Dupe*)

Eddie Annual awards given for editing by the American Cinema Editors.
Edge numbers (*See Key Numbers*)

Edit To select and organize all the takes provided by the director while using a creative and technical expertise to tell the story as best as possible.

Editor The person responsible for the editing. This term usually refers to a *film* or *videotape editor* but may also refer to *trailer editor, music editor, sound editor, ADR editor, foley editor, dialogue editor,* and so forth.

Effects, EFX, FX (*See Optical effects* and *Sound effects*)

Emmy Annual awards given by the Academy of Television Arts and Sciences.

Emulsion The face-up, dull side of the film as compared to the shiny, celluloid surface on the reverse side. (*See Cell*)

End markers, end slates, or **end sticks** Clapsticks placed at the end of a take usually upside down.

End cues (*See Changeover cues*)

End pop or **tail pop** A one-frame sound tone or *beep* placed in the tail leader usually three feet from the end of the track. This with a head pop enables the negative cutter to verify the sync between the optical track and cut negative picture before a composite print is made. (*See Head pop*)

End syncs White grease pencil marks placed by the assistant editor precisely opposite each other at the end of the last take of a picture and track daily reel so that the apprentice can verify that the reel had been coded correctly.

Fade in An optical effect in which a scene gradually appears.

Fade out An optical effect in which a scene gradually disappears.

Feet or **footage** The measurement of film. In 35mm film one foot contains 16 frames. 16mm film has 40 frames per foot.

Female flange A metal or plastic disk that has a smooth appendage upon which to roll up the film without a core. (*See Flange*)

Fill Temporary background sound or leader used to replace any disturbing production sound in the track or a scene that has been shot silent.

Film horse or **leader stanchion** As a self-standing wooden apparatus or as a metal attachment to the left rewind it is used to supplement the rewind by holding and unwinding film—usually leader.

Film library A specialized house or a department in a film company that catalogues and files material that can be reused. (*See Stock*)

Fine grain A master positive stock made from the original B&W negative and used in the duplicating and optical processes. (*See Interpositive*)

Flange A metal or plastic disk that fits onto a rewind shaft and is used to roll up film. (*See also Adaptor* and *Female* and *Male flange*)

Flatbed (*See Horizontal editing machine*)

Flip An optical that in only eight frames creates a fast turning-over effect by literally squeezing out the outgoing scene while in reverse manner bringing in the incoming scene.

Flop or **flopover** Optically reversing the cell side with the emulsion side of a scene thus reversing the action.

Focus chart Film leader containing a diagram of vertical and horizontal lines placed ahead of the first take on each daily reel to assist the projectionist in establishing ideal focus.

Foley The process of creating sound effects especially *footsteps* on a sound, or foley, stage.

Footage chart A handy form used to estimate footages and corresponding running times.

Frame (*noun*) Each individual picture representing one-sixteenth of a foot and containing four sprocket holes on each side. (*verb*) To adjust the picture so that each frame will be exactly within the aperture of the viewing apparatus. If the *frame line* separating each frame is visible on the screen, the picture is *out of frame,* and when it is corrected or *framed,* the picture is *in frame.*

Frame line The visible line separating each frame of picture. (*See Frame*)

Frame markers, pins, or **tits** Small dark triangular marks on the left edge of some film stock directly aligned with the frame lines to assist in locating them.

Framer (a) A rotating disk on the synchronizer numbered 15 frames with the sixteenth frame as the *0* frame. (b) A mechanism on the moviola used to adjust the framing of the viewing aperture.

Frame ruler Any type of measurement guide for one or two feet provided at the coding machine.

Freeze frame, hold frame, or **stop frame** A selected frame of picture optically repeated for any desired length.

Full coat Track stock fully coated on the oxide side as compared to the *single stripe* mag track.

Full in The precise frame at which a fade in or the *B* side of an optical is completed.

Gang (*See Synchronizer*)

Garbage Commonly-used slang term for *network format* material required for television, excluding story action.

Generation As in second, third, and so forth generation referring to the number of successive negative duplications from the original negative.

Goodies Usable production track material saved by the editor for sound effects.

Handle The additional one or more feet ordered at the head and tail ends of an optical count.

Head or **heads up** The beginning of a reel or roll of picture or track.

Head pop A one-frame sound tone placed at the three-foot track academy leader. (*See End pop*)

Hold frame (*See Freeze frame*)

Horizontal editing machine or **table** or **flatbed** A counter-style viewing apparatus containing source plates on the left and take-up plates on the right for picture and track rolls or reels that can be run in sync or separately by control switches. There are one or several screens at the rear of the table for viewing. The models principally used in Hollywood are *Kem, Moviola,* and *Steenbeck.*

Hot splicer A large, heavy piece of electric equipment used to splice two pieces of film together with cement. Primarily used in negative cutting. (*See Butt splicer*)

In frame (*See Frame*)

Insert A close shot of some particular object or piece of action, e.g., a newspaper headline or the trigger of a rifle being squeezed.

Inside counts Principal method of ordering opticals—the *full in* count being the last frame of the effect and the *start out* count being the first frame. (*See Outside counts*)

Inside optical Limiting the ordering of an optical with a handle just past the full in count and just ahead of the start out count. (*See Outside optical*)

Interlock The connector on the drive shaft of the moviola that, when joined, or *in* interlock, enables picture and track to run simultaneously, and when released, or *out of* interlock, to run separately.

Internegative (interneg) A second or lower generation color negative processed from an *interpositive.* Also loosely referred to as a *dupe negative.*

Interpositive (I.P.) A special color print made from the original negative and used to make *internegatives.*

Irises *Iris in* is optically beginning a new scene within an expanding circle. *Iris out* is optically ending a scene within a diminishing circle.

Jump cut Usually the accidental elimination of one or more frames within a cut thus negating the mandatory 16 frames per foot and jarring the viewer. When it is done intentionally in certain situations, however, the jump cut may not be perceived and can speed up the action.

Kem® Proprietary name for a popular flatbed; also used as a generic term for the picture and track rolls used on any horizontal editing machine, i.e. *kem rolls.*

Keying Listing the key numbers by the negative cutter from the cut picture so that the corresponding negatives might be acquired and organized prior to cutting negative.

Key numbers or **negative key numbers** The consecutive numbering of negative stock at one-foot intervals. These numbers are repeated on the positive stock and serve as identification of the material. (*See Code numbers*)

Lab report Form sent with each roll of dailies listing all takes included and noting any irregularities or other pertinent information.

Lay back A videotape on-line session during which the dubbed 24-track master is married to the one-inch video master.

Lay down A videotape on-line session during which the edited picture, titles, opticals, and production track are transferred onto masters.

Leader Blank colored film used at the head and tail of a picture to identify and protect it or used as fill in the track. Discarded B&W reversal prints are also used as sound leader.

Library stock report An accounting of the exact footage of any *outside* stock used submitted by the assistant editor to the company stock library.

Lift (*noun*) A portion of a scene or an entire scene that has been removed from the picture during editing. (*verb*) To remove.

Lined script The shooting script that the script supervisor sends to the editor. Lines down each page identify the takes involved, where they begin and end, and which angles have on-screen dialogue or action.

Line up (a) To put in sequence as in putting reels in the order desired for projection or putting daily rolls in scene sequence or lined script sequence for editing. (b) To align two or more pieces of film in a synchronizer or viewer by start marks, clapsticks, or code numbers.

Lip sync Matching lip movements and dialogue or any other action and sound to achieve sync.

Locked in (a) Signifying that the editing of the picture is finished with no further editing changes anticipated. (b) Expression used by a picture editor when the director has failed to shoot any alternative angles that might be used.

Logo The filmed emblem of a motion picture company on its product.

Loop (*noun*) (a) A section of production dialogue joined beginning to end to form a continuous repetitive track in *looping*. (b) The short slack of picture required ahead of the viewing aperture when you are threading up for projection. (*verb*) Referring to an obsolete method of preparing track for the rerecording of dialogue in sync with picture. (*See ADR*)

Mag or **magnetic track** Film stock having a sensitive iron oxide coating for the reproduction of sound.

Mag down The oxide or dull side of the track face down.

Mag up The oxide or dull side of the track face up.

Male flange A metal or plastic disk that has a center appendage that holds a core upon which to roll up the film. (*See Flange*)

Master The full scene or shot showing the geographical relationship of the performers and the locale.

Match cut (*noun*) The point at which two different shots are joined together resulting in continuous smooth action. (*verb*) Editing such a cut.

Matching shot One angle as related to another angle.

Matte shot Photographing a scene with a section of it blacked out—or to be blacked out or covered later—so that it might be filled in with a second shot or painted art work.

Misslate A misnumbered scene or slate number on the clapsticks.

Mix (*See Dub*)

Mixer The sound person or recorder on the set during shooting; also a sound technician involved in the dubbing of a show.

Montage A series of fast cuts or of various optical effects such as superimposures to convey the transition of time, action, or events in minimal screening time.

MOS Abbreviation for picture shot without sound; a silent (*Sil*) picture.

Move-in Optically moving in and ending as a blowup of a subject. (*See Blowup*)

Moviola® Proprietary name of a horizontal editing table and for the most popular American upright machine; also used as a generic term for any upright viewer.

Mutliple cameras More than one camera shooting a take. Each camera is identified as *A* camera, *B* camera, *C* camera, and so forth.

Music code In addition to daily code numbers recoding *music takes* usually in *yellow* in sync with a music playback.

Music cues The precise footages at which each section of music in the film begins and ends. These are logged on a *music cue sheet* by the *music editor.*

Music editor The editor who is responsible for providing the composer with *music timing sheets* and preparing music tracks and *music cue sheets* for dubbing.

Music playback (*See Playback*)

Music timing sheets A detailed record of the running time, dialogue, and action of each music cue for the composer.

Negative cutter The person responsible for matching original negative, cut for cut, to the final edited work print.

Negative key numbers (*See Key numbers*)

Network format Television directive specifying the makeup of a show. (*See Garbage*)

No ann Abbreviation for *no announcement* of the scene or slate on a sound take in the dailies.

Numbering block A round metal apparatus on a coding machine that provides a variety of numbers and letters for coding.

Numerical slate coding This coding system is used when takes are shot by slate numbers rather than by scene and take numbers, and each slate is identified by its code number. This is accomplished on the American coding machine by replacing the numbering block with a special block that prints first the take number, 1 to 0 (for 10); then the slate number with one to three digits using first a numeral and then one or two letters, A through K except I, each representing numerals, 1 through 0 (A = 1, B = 2, and so forth); and finally, the footage is represented by three numerals usually beginning with 000. Examples: 0 8 000 is slate 8, take 10; 6 A2 115 is 115 feet into slate 12, take 6.

Off camera or **off screen** Action or dialogue that is occurring outside of camera range.

On camera or **on screen** All action or dialogue that is occurring within camera range.

One-light Direction on a camera report that the lab develop the film using a constant or one-light setting.

One-liners (*See Script notes*)

On footage Meaning that the picture is at the exact footage required. (*See Action footage* and *Locked in*)

Optical effects A variety of photographic impressions created with production material and other film in an optical printer to serve as transitions in time, locale, and action to create a dramatic or comedic effect, or to correct some error in the original production.

Optical track Photographic sound used in the editing of film during the first three decades of *talkies.* It is now used only as a final dubbed track that is synced to the cut negative picture and appears on the left side of the composite print as a band of visible striations.

Oscar A statuette awarded annually by the Motion Picture Academy of Arts and Sciences.

Out of frame (*See Frame*)

Outs Those daily takes that an editor has left intact.

Outside counts Rarely used method of ordering opticals—the *full in* count being the first frame immediately following the last frame of the optical and the *start out* being the frame immediately preceding the first frame of the optical. (*See Inside counts*)

Outside optical Sometimes referred to as a *full* optical—the ordering of the entire cut containing an effect. (*See Inside optical*)

Outside stock Stock film obtained from a source outside of your film company for use in your company's film. (*See Film library*)

Outtakes Those takes that a director has not ordered printed. (*See B Negatives*)

Over and over Standard winding of film off the top of the source reel onto the top side of the take-up reel; winding *straight across.* Film will remain the same side up. (*See Under and under*)

Over and under Winding of film off the top of the source reel onto the bottom or underside of the take-up reel thus reversing the side up of the film. (*See Under and over*)

Overcrank Increasing the camera speed over the normal rate of 24 frames per second. The result is in slower motion on film. (*See Undercrank*)

Overlap Usually refers to the extension of sound so that it continues simultaneously with incoming action.

Paper or **paper off** To mark a specific section of film in a reel or roll by folding a piece of paper around the film or by spotting the section with a piece of tape.

Paper to paper Ordering only a section of a take with beginning and end key numbers as opposed to ordering the entire take.

Perforations or **perfs** (*See Sprockets* or *sprocket holes*)

Personal sync mark (PSM) Used as a reference point on picture and track in the syncing of dailies.

Pick up (P.U.) A portion of a take shot to supplement the original shooting of the same angle. Also the shooting of an additional take or scene after the principal photography has occurred.

Picture The visual film but also includes the sound when referring to the complete film.

Pins (*See Frame markers*)

Playback A recording that serves during shooting as a sync guide for singers, dancers, and musicians.

Pop (*noun*) (a) An unwanted sound in the track usually at a defective splice. (b) A one-frame sound tone or *beep* placed in the head and tail academy track leaders. (*See End pop* and *Head pop*) (*verb*) To mark the sounds of the clapsticks on the daily tracks of each take in preparation for their syncing up with the picture.

Pop on or **pop off** Beginning or ending picture, titles, or credits without a *fade in,* a *fade out,* or any other effect.

Positive or **positive print** A color or black and white visual film developed from the negative. (*See Print*)

Pre-dub A *temporary* rerecording of two or more tracks onto one track.

Preview coding The coding of the composite picture or work picture and any necessary tracks when additional editing is required after a preview.

Print (*noun*) A photographic copy of the picture made from the negative. (*See Work print*) (*verb*) To develop a copy of the picture from the negative.

Promo Important segments of a film edited together for promotional purposes. (*See Public service*)

Public service Additional time, ranging from ten seconds to one minute, provided by a network or station for promotional material considered to be in the public interest.

Pull up The minimal 28-frame clearance of any dialogue provided by most editors at the heads of each reel to enable sound to be in sync with picture when it becomes optical track on the composite print.

Punch (*noun*) A metal device used to notch film that requires numerical slate coding. (*verb*) To notch the film.

Reconstitute Use of leader on picture or track rolls to maintain sync during editing that is being done on a flatbed.

Red line Use of a red marker to line the right, clear side of a take such as stock as a reminder that if it is used in the picture it will have to be duped.

Reel rack A self-standing metal tiered framework for reels that is also used (cautiously) to file trim boxes.

Registration fine grain/interpositive Printed frame by frame for optimum quality and maximum stability. This is required for certain opticals and titles.

Release print The final composite print used for theatrical exhibition.

Reposition Optically adjusting the frame image and usually used in conjunction with a move-in or blowup.

Reprise Generally used after the first episode of a miniseries. Beginning each subsequent episode with a repetition of the major story developments that preceded it.

Retake The reshooting of a previously shot scene. (*See Pickup*)

Rewind (*noun*) (a) A device attached to both sides of an editing bench that is used to wind film. (b) The *moviola rewind* is a female flange-type attachment on the side of the moviola with which to rewind a roll of film after running.

RGB Appears on the lab report and refers to the red, green, and blue colors of light necessary to produce a color print.

Room tone The background sound in a room of a home, in an office, and so forth. (*See Ambience*)

Rough cut Often misused term for a *first cut* or *director's cut.*

Running A screening

Rushes (*See Dailies*)

Scale The minimal salary for each editorial category as negotiated by the signator companies and the Guild.

Scene Literally a part of an act or sequence having continuous action in a single locale but often diversely used to refer to an entire *sequence* or a single *take*.

Scene numbers As designated by the script supervisor when she is lining the script and thereafter used in post-production. Her scene numbers do not always correspond exactly with the scene numbers in the script.

Scoring session Musicians on a sound stage recording the music composed for a film.

Screening Viewing a film in a projection room or theater.

Scribe (*noun*) A metal, pencil-shaped instrument used to write on film. (*verb*) To write on film.

Script notes A daily listing of the *circled* takes by the script supervisor with a brief description of each take. Also called *one-liners*.

Script supervisor Member of the crew who, besides being responsible for the script notes and lined script, assists the director during shooting by verifying the dialogue and making certain many other production and technical details are accurate.

Sequence A series of incidents or scenes unified by time, place, and related action. (*See Scene*)

Set up The positioning of the camera and lighting. Any change constitutes a new angle or shot.

Shaft The spoke of a rewind that can hold four to six reels depending on its length.

Short ends or **waste** Negative on a camera roll not used.

Short trims The editor's trim box for any trim less than a foot in length.

Shot The elemental component of a scene distinguished by the camera angle and the distance between the camera and the subject matter. (*See Take*)

Sil Abbreviation for silent. (*See MOS*)

Single stripe Magnetic work track with a single channel for the sound.

Skip frame or **skip printing** Optically eliminating frames within a cut to speed up the action. (*See Undercrank*)

Slate (*See Clapsticks*)

Sound editor Preferred title for editor who cuts sound effects.

Sound effects Implementation of the original production track and substitution with various sounds taken from a sound effects library to enhance the film. Also includes *ADR, dialogue, editing,* and *foley.* Not to be confused with *special effects.*

Sound leader (*See Leader*)

Sound report A daily record submitted by the mixer or recorder on the set indicating all the takes shot that day including the circled takes and other pertinent information.

Sound tone An audio signal of a single frequency in the dailies.

Sound transfer report A listing of all the circled takes that have been transferred by the transfer room of a sound department or house.

Splice (*noun*) The place where two pieces of picture or track have been joined. (*verb*) To join two pieces of film.

Split reel A hybrid device similar to both a flange and a reel for winding film onto a core. It is metal and the back side screws into the male-like flange side that holds the core onto which the film is wound.

Split screen One or more separate images projected optically onto a single frame.

Spot code The coding of picture or track to begin at a precise point with an exact code number.

Spotting session (a) A screening of the edited picture for *sound effects* to discuss where sound effects, foley, or ADR is required. (b) A screening of the film for the *composer* to discuss the location of music cues and the kind of music to be used. (*See Music cues*)

Sprockets or **sprocket holes** Perforations (perfs) on the edge of the film for threading through a camera, projector, or editing equipment. 35mm film has four evenly spaced sprocket holes on both sides of a frame. It is referred to as a four-sprocket frame.

Standard coding (*See American coding*)

Start or **start mark** The frame in the head leader usually identified with an *X* so that picture and track might be threaded up in sync.

Start out The exact frame that begins the optical effect of an outgoing scene.

Stock Scenes that might be reused such as establishing shots, car runbys, disasters, and so forth.

Stop and go A screening for the producer or director that is halted whenever some phase of editing has to be discussed.

Straight cut Cutting directly from one scene to another without any optical effect for the transition of time or place.

Straight run through An uninterrupted screening of a film.

Superimposure or **super** Optically overlapping two or more scenes onto one piece of film.

Sweeten Term used in the dubbing process meaning to add to something in a particular area.

Swivel base A movable base on the left rewind for more efficient running of the film from the bench into the moviola.

Sync or **synchronize** To match picture and track.

Synchronizer A device with two or more sprocketed *gangs* to hold picture and track in sync. A sound head with amplifier is attached for auditing tracks, and there is a footage counter to register the measurement of the film.

Tab A small card attached to each daily take by the assistant editor as the dailies are broken down. It should contain the slate identification, brief description, beginning and end code numbers, and any other pertinent information.

Tag Brief concluding scene in a few television series following the climax or last act of an episode usually with an intervening commercial break.

Tail, tails out, or **tails up** The end of a reel, roll, or piece of film.

Tail-to-head reverse or **reversal** Optically reverses the action of a cut.

Take Each shooting of the same angle of a scene until the director is satisfied. They are consecutively numbered, and those the director wishes printed are circled on all the reports.

Take-up moviola Editing machine with arms attached to hold reels so film being viewed can simultaneously be rewound as in a projector.

Taking counts Ordering opticals.
Teaser A short segment containing cuts from several scenes of the film placed before the beginning of a television show to entice the viewer.
Templates Sample film or leader used as guides for ordering complicated opticals or for measuring end cues.
Title copy A listing of the picture title and of all screen credits for main title, opening credits, and end credits.
Tits (*See Frame markers*)
Three-gang or **three-way** (*See Synchronizer*)
Three-stripe A track with three channels.
Tracking (a) The re-editing of previously scored music from a music library. (b) Shooting with a moving camera following or preceding the action. Also called *traveling* or *trucking*.
Track leader (*See Leader*)
Trailer Several of the most entertaining segments of a film edited together for a brief sequence for the purpose of advertising the film.
Trim (*noun*) A section of a take not used in the edited film. (*verb*) To edit out an unnecessary part of the action.
Trim bin A large metal container on wheels fitted with an inner bag and used as a temporary receptacle for film trims during editing.
Trim boxes Cartons used to hold trims, rolls, or outs and filed in numerical order either by scene or slate numbers or by code numbers.

Under and over Winding film off the bottom of the source reel onto the top of the take-up reel, thus reversing the side up of the film.
Under and under Winding film off the bottom of the source reel onto the bottom of the take-up reel. The film will remain the same side up.
Undercrank Decreasing the normal camera speed of 24 frames per second for faster film motion. (*See Overcrank*)
Units The various reels *built* by the sound effects editors for each picture reel.

Variable speed The use of the right pedal of the moviola to vary the speed of viewing the film according to the amount of pressure exerted on the pedal. (*See Constant speed*)
Vignette (a) A brief scene or sequence meant to imply a broader action or series of events. (b) An optical effect created with a curlicued type of matte.

Webril wipes Disposable cotton wipes used in the cutting room to clean off dust particles from the film.
Wild lines or **wild track** Dialogue or background sound recorded without picture to be synced with the corresponding picture later.
Wind down (*noun*) Picture and/or track wound down for viewing with temporary start marks about 10 feet before the requested section. (*verb*) Rewinding picture and/or track to a certain place.

Wipe The optical transition from one scene to another by which the incoming *B* scene as it comes on full screen *wipes* the outgoing *A* scene off the screen.

Work picture or **work print (W/P)** Usually refers to the complete film including the sound during editing. (*See Picture*)

Work track (W/T) The sound film during editing.

FURTHER READING

Anderson, Gary H. *Video Editing and Post-Production: A Professional Guide.* White Plains, N.Y.: Knowledge Industry Publications, 1984.

Balmuth, Bernard. *The Language of the Cutting Room.* North Hollywood, Calif.: Rosallen Publications, 1978.

Browne, Steve. *Video Editing: A Post-Production Primer.* Boston: Focal Press, 1989.

Brownlow, Kevin. *The Parade's Gone By.* New York: Bonanza Books, 1968.

Burder, John. *The Technique of Editing 16mm Films.* Boston: Focal Press, 1988.

Case, Dominic. *Motion Picture Film Processing.* Boston: Focal Press, 1985.

Crittenden, Roger. *Film Editing.* London: Thames and Hudson, 1981.

Dmytryk, Edward. *On Film Editing.* Boston: Focal Press, 1984.

Goodman, Ezra. *The Fifty-Year Decline and Fall of Hollywood.* New York: Simon and Schuster, 1961.

Happe, L. Bernard. *Your Film and the Lab.* Boston: Focal Press, 1983.

Jacobs, Lewis. *The Rise of the American Film.* New York: Harcourt, Brace, 1939.

Lustig, Milton. *Music Editing for Motion Pictures.* New York: Hasting House, Publishers, 1980.

Miller, Pat P. *Script Supervising and Film Continuity.* Boston, Focal Press, 1986.

Pacific Coast Studio Directory. Hollywood, Calif.: Published quarterly. Harry C. Reitz, Publisher; Jack C. Reitz, Editor.

Reisz, Karel, and Gavin Millar. *The Technique of Film Editing,* 2nd ed. Boston: Focal Press, 1968.

Rosenblum, Ralph, and Robert Kuren. *When the Shooting Stops . . . The Cutting Begins.* New York: Viking Press, 1979.

Schneider, Arthur. *Electronic Post-Production and Videotape Editing.* Boston: Focal Press, 1989.

Walter, Ernest. *The Technique of the Cutting Room,* 3rd ed. Boston: Focal Press, 1969.

INDEX